INDIA: ECONOMIC, POLITICAL AND SOCIAL ISSUES

SCIENTIFIC RESEARCH AND MANUFACTURING

POTENTIAL INDIA GROWTH STORY

India: Economic, Political and Social Issues

Additional books and e-books in this series can be found on Nova's website under the Series tab.

India: Economic, Political and Social Issues

Scientific Research and Manufacturing

Potential India Growth Story

B. M. Suri

nova science publishers
New York

NOTICE TO THE READER

Library of Congress Cataloging-in-Publication Data

ISBN: 978-1-53615-230-2

Published by Nova Science Publishers, Inc. † New York

Dedicated to my loving parents

CONTENTS

FOREWORD

The world is witnessing a dramatic rise of disruptive technologies like artificial intelligence, machine learning, big data analytics, adaptive manufacturing etc. which will have tremendous influence on all aspects of our lives. Their impact on jobs, economic growth, social behavior, government policy, legal jurisprudence and sustainable development will be significant and unpredictable.

As has been the trend in the past, these technologies are usually developed as owned by western countries including the multinational corporation. Due to the lack of adequate Science, Technologies, Engineering and Mathematics (STEM) education, R&D ecosystem, resources and weak governance, developing countries find it extremely difficult to develop new technologies. These technologies and the products are then sold to the developing countries which then become dependent upon the developed countries for their technology needs. Due to the existence of numerous deliberately crafted barriers, rarely are these technologies transferred to developing countries. There is a risk that the prevalent technology divide between developed and developing countries is likely to deepen further. This will create developmental disparities which in turn would breed tension. The impact of these technologies on the environment is also a matter for concern.

In the last few decades, India has registered impressive economic growth and has become the sixth largest economy in the world. India is using technologies with great effect to leapfrog the various stages of development and improve delivery of services and bring efficiency in governance. Its "Make in India" initiative is aimed at making it a manufacturing hub of high technologies. The 'Make in India' program is being dovetailed into a "designed, developed and Made in India". However, this progression will not be easy as it will require a fundamental reform of the education system, R&D institutions, manufacturing practices and Intellectual Property Rights (IPR) framework. Such a comprehensive reform needs a broad vision, policy and the matching institutions.

The present book by Prof BM Suri deals with these issues in a comprehensive way. The author had a long carrier as a scientist and technologist in the Bhabha Atomic Research Centre. The book, written on the basis of the author's four decade of experience, tells the reader in an authentic way what it takes for a nation to become a leader in high technology. It also provides an insight into the deficiencies of the S&T Indian ecosystem and how these can be redressed. The author provides useful advice to the policy makers. He recommends the formation of an apex body NSTIS in order to make a suitable policy, monitor its implementation and ensure coordination amongst diverse scientific, technological, educational and manufacturing institutions.

This pioneering work will stimulate debate on how India should harness technologies for the wellbeing of its people.

Arvind Gupta
Director,
Vivekananda International Foundation, New Delhi, and
Former Dy. National Security Adviser,
Government of India

New Delhi
01 October 2018

PREFACE

India and many other populous developing countries are facing a severe crisis related to joblessness. About a decade ago, the growth of jobs in India could almost keep pace with the jobseekers, mainly because the Indian IT sector was growing rapidly. However, that is not the case now. With the advent of technologies like artificial intelligence, machine learning and robotics, internationally, the number of jobs has started declining. This has impacted the Indian IT sector and consequently the scope of related employment. These developments have come at a time when India is struggling hard to provide jobs to its millions who are moving away from agriculture. Additionally, a large part of this new workforce is unemployable for the new jobs being created because of these new technologies. This is one of the reasons why industrial manufacturing has acquired great significance in the eyes of Indian policymakers. Contrary to the general trends, India's service sector has grown much faster than its manufacturing sector. This has happened in the backdrop of several of India's Asian neighbours progressing quite fast in manufacturing. They have become part of global supply chains. This has happened due to the strategy adopted by manufacturers in developed countries to shift and fragment their operations to minimise the cost of production. Many of the developing countries made use of this opportunity by providing favourable conditions for such manufacturing. India, being a latecomer, may find it difficult to beat its competitors at this stage. All these factors have to be viewed together with the continuously reduced competitiveness of Indian manufacturing exports. These problems collectively pose a serious challenge to India.

Developed countries do not face serious problems of unemployment, despite a major shift of their manufacturing to developing countries. Some of the reasons behind this are that they bank heavily on high technology-based manufacturing and services and have much smaller populations. Such manufacturing and services are backed by high-end scientific research and technology development. India too, in principle, has the potential to take a similar route and transform these challenges into opportunities. Its biggest assets are its internal demand and talent. Both these assets are presently being made good use of by external agencies. Some of the factors that deserve consideration are technology-based imports into India, foreign countries taking away its talent, and in recent times, multinational R&D centres employing its talent. India has yet to fully harness its own talent (provided it retains them well!) for successfully attaining its own national objectives, including meeting the twin challenges of unemployment and disruptive technologies.

India, in the year 2014, launched mission 'Make in India' to boost its falling manufacturing industries. The mission focused on 25 sectors, and is being run by the Ministry of Commerce and Industry. The emphasis is on policies to ease investments. In fact, a major yardstick of progress is foreign direct investment (FDI). The mission statement does include the word innovation, but not much beyond. There is no mention of linking this mission with indigenous scientific research and technology development. This is indicative of poor expectations of contributions of Indian R&D to domestic industrial manufacturing. It seems the mission lays excessive emphasis on capital over human capital.

This book is a modest attempt to highlight how indigenous R&D can be a tool of economic growth (and quality jobs) instead of mostly serving national strategic needs, as it presently is. This is how countries like Japan, Korea and Israel have progressed with the number of scientific researchers in the range of few thousands per million of the population, as compared to a meagre 218 for India. This is when the demand for high technology products is very high within the country. India's defence and electronic product imports are expected to run into hundreds of billion dollars in the next few years. India's R&D sector, together with manufacturing, is not yet able to fulfill these expectations. Most of these strategic requirements related to defence, atomic energy, or space exploration rest on human resources from STEM HE institutions. India's STEM HE is not yet able to supply the requisite quality and quantity of human resources to these sectors. There are small numbers of elite institutions producing disproportionately low number of trained workforce and a substantial fraction is lost from the STEM ecosystem and even from the country, as the system is not able to absorb them. India's mass institutions are extremely impoverished in terms of research infrastructure to be able to fill the gap. A significant question in this context is about how research directions in India contribute to its manufacturing industries and in building the industrial base. It seems there is a poor connection between Indian R&D and manufacturing industries. There is a dominance of culture of paper publishing over that of patents and product development, and R&D directions are not tuned to the demands of hi-tech industry. This is while a large number of MNC R&D centres in India are able to produce a good number of commercially viable patents by using local Indian talent. R&D directions yoked to international trends and the frontline are likely to deprive innovation of its genuine glory and worth, except winning the certification of international 'also ran' category. It is important that researchers and research managers realise the worth of patents and domestic demands. Performance audit mechanisms of Indian R&D may need to be reviewed for better tuning to high technology development needs.

Indian researchers, educators, and innovators have several challenges to confront in the context of hi-tech manufacturing. They have to reconcile the contradictory demands of improving industrial productivity and yet maximising employment. There is need to build world-class institutions and develop the ability to retain talent in the face of competition from developed countries. They have to develop quality products, but with favourable economics and improved accessibility. All of this has to be met in the face of several disruptive technologies on the horizon. These are the challenges faced by scientists and technologists. Does India, as a nation, have an appropriate response system? This author, during his more than three decades of interaction with Indian science and technology ecosystem, observed several opportunities where Indian R&D could have usefully contributed to industrial manufacturing. Always, the bus was missed because of issues related to demand economics, last mile shortcomings, limitations of mandate, and poor entrepreneurial spirit among the scientific community, apart from many other such factors. Gradually, the relevant competence developed was also frittered away. In all, hi-tech manufacturing, fast response, and first mover advantage is imperative. Overcoming barriers in translation of R&D to hi-tech manufacturing is a challenge that requires response from beyond the R&D sector.

India must make use of the experience of many of the developed countries and realise that R&D and technology based young startups have the potential to provide a large number of new jobs, with an essential aspect being the wide dissemination of pertinent knowledge base and skillsets. This may provide the hope of harnessing the dividend of India's demography, while it confronts the challenges of futuristic disruptive technologies. These and many such challenges are commonly faced by many developing countries. India may do well to take the initiative to bring several of these countries together for this purpose, similar to the International Solar Alliance.

This book addresses a broad spectrum of audiences that includes academicians, scientists, technologists, innovators, entrepreneurs, research managers, policymakers, and business school mentors. The foremost objective of this work is to alert the policymakers to the possibilities and pitfalls in the context of scientific research, industrial manufacturing, and disruptive technologies of the future. It is aimed to sensitise scientists and technologists to the significance of patents, both in individual as well as national context. The book pleads with research managers to facilitate spinning off technology-based startups and expects teachers and mentors in STEM HE to inculcate the spirit of innovation and enterprise in young students. It suggests how B Schools can mentor and guide Indian researchers into the world of enterprise and facilitate marketing and branding of Indian high technology products. It also highlights pertinent issues with examples from technologically developed countries to point out the strategic significance of STEM HE, scientific research, and relevant action points in local context. The book avoids the use of technical jargon so that it is easily understood by the public at large. The author will be grateful for constructive criticism that aids the requisite debate on the issues raised in this book.

B. M. Suri
Mumbai, India

ACKNOWLEDGMENTS

I thank Dr. Arvind Gupta, who set me thinking about answers to several provoking questions about effectiveness of Indian scientific research in the context of manufacturing. Dr Gupta made several useful suggestions regarding the manuscript. I thank Prof. Amitava Sengupta and Dr. Gulshan Rai for patiently going through the full length of the manuscript and making valuable suggestions. I am grateful to Prof. K. K Agarwal, Dr. P. A. Naik and Prof. Arun Grover for showing interest in the subjectt and sharing their valuable feedback. I express my gratitude to Ms Smita Purushottam and the SITARA discussion group for enlightening discussions on several relevant subjects. I am grateful to Prof. Amitabh Joshi, Dr. S. K. H Auluck and Prof. Manas Mukherjee for critically assessing preliminary versions of this work.

I owe my gratitude to a number of people that includes my teachers, family and friends for their constant support and encouragement. I thank my wife Ritu, without whose support I could not have completed this work. I am particularly grateful to Dr. Anil Pulhani and Mr. Girish Zende for going through the manuscript in detail and formatting it. I thank Sheetal and Piyush Suri for meticulously finalising and formatting all the figures mentioned in the work and obtaining copyright permissions for the same. I am grateful to Sucharita Sengupta Suri (seekred.com) for designing artwork for the cover page of the book. I thank all my family members, who always supported me in my endevaours. I thank Mr. Avneep Dhingra for editing the book in full and bringing it to final shape. Finally, I thank my friends and colleagues with whom I might have discussed several issues (and benefitted in the process) relevant to this book.

ABBREVIATIONS

AEE	Atomic Energy Establishment
AICTE	All India Council of Technical Education
AIIMS	All India Institutes of Medical Sciences
BHEL	Bharat Heavy Electrical Limited
BPO	Business Process Outsourcing
BRICS	Brazil Russia India China and South Africa
BRNS	Board of Research in Nuclear Sciences
CAG	Comptroller & Auditor General
C-DOT	Centre for Development of Telematics
CDRI	Central Drug Research Institute
CERN	Centre for Experimental Research in Nuclear Physics
CKD	Completely Knocked Down
CLW	Chitranjan Locomotive Works
CSIR	Council of Scientific & Industrial Research
CSR	Corporate Social Responsibility
DAE	Department of Atomic Energy
DIY	Do it yourself
DOS	Department of Space
DRDO	Defence Research &Development Organization
DST	Department of Science & Technology
DTTI	Defence Trade and Technology Initiative

FBAR	Film Bulk Acoustic Resonator
FDI	Foreign Direct Investment
FICCI	Federation of Indian Chambers of Commerce and Industry
GDP	Gross domestic Product
GERD	Gross Expenditure on R&D
GPS	Global Positioning Sensors
GSLV	Geosynchronous Satellite Launch Vehicle
GSTN	Goods & Services Tax Network
GW	Giga Watt
HAL	Hindustan Aeronautics Laboratory
HEI	Higher education Institutes
HMT	Hindustan Machine Tools
IAF	Indian Air Force
ICAR	Indian Council of Agricultural Research
ICMR	Indian Council of Medical Research
ICT	Information and Communication Technologies
IICT	Indian Institute of Chemical Technology
IIM	Indian Institutes of Management
IISER	Institutes of Science Education and Research
IISc	Indian Institute of Science
IIT	Indian Institutes of Technology
IP	Intellectual property
IPR	Indian Patent Rights
ISA	International Solar Alliance
IT	Information Technology
ITER	International Tokomak Experimental Reactor
KBC	Knowledge Based Capital
KBE	Knowledge Based Economy
KI	Knowledge Intensive
LCA	Light Combat Aircraft
MBT	Main Battle Tank
MIT	Massachusetts Institute of Technology
MNC	Multi National Corporation
MNRE	Ministry of New and Renewable Energy Sources

MOOC	Massive Online Open Courses
MOU	Memorandum of Understanding
MSME	Micro Small and Medium Enterprises
MTCR	Missile Technology Control Regime
NCPVR	National Council of Photovoltaic Research
NCL	National Chemical Laboratory
NCR	National Capital Region
NGT	National Green Tribunal
NIS	National Innovation System
NPL	National Physical Laboratory
NPT	Non-Proliferation Treaty
NRI	Non-Resident Indians
NSTIA or NSTIS	National Science Technology and Innovation Authority or System
NTPC	National Thermal Power Corporation
NZL	New Zealand
OECD	Organization of Economic Cooperation & Development
OEM	Original Equipment Manufacturers
O&M	Operations and Maintenance
PHWR	Pressurised Heavy Water Reactors
PPP	Purchasing Power Parity
PRI	Public Research Institutes
PSU	Public Sector Undertakings
R&D	Research and Development
SAC	Scientific Advisory Committee
SME	Small and Medium Enterprises
SKD	Semi Knocked Down
STEM	Science Technology Engineering and Maths
STEM HE	Science Technology Engineering and Maths higher education
S&T	Science and Technology
TRIPS	Trade Related Intellectual Property Rights

TIFR	Tata Institute of Fundamental Research
UGC	University Grants Commission
UNICEF	United Nations International Children Emergency Fund
WIPO	World Intellectual Property Organization
WHO	World Health Organization

NAMES OF COUNTRIES (USED IN FIGURES)

CAN	Canada
ISR	Israel
IRL	Ireland
LUX	Luxemburg
AUS	Australia
GBR	Great Britain
NZL	Newzeland
POL	Poland
EST	Estonia
CHE	Chile
DNK	Denmark
SWE	Sweden
LTU	Lithuania
JPN	Japan
HUN	Hungary
ISL	Israel
BEL	Belgium
FIN	Finland
FRA	France
LVA	Latvia
SVK	Slovakia
KOR	Korea
PRT	Portugal
NLD	Netherland
CZE	Czech Republic

ESP	Spain
TVR	Turkey
DEU	Germany
AUT	Austria
HRY	Croatia
GRC	Greece

Chapter 1

INTRODUCTION

Sustainable development as a concept has caught the attention and become significant to a large number of people worldwide in recent times. Sustainable development, besides other attributes, is a form of growth with a more predictable pattern. There could be several factors which may impact global sustainable development in the near future. These include threat to climate, onward march of several new disruptive technologies, unfavourable demography, large-scale immigration and shortage of talent required to handle frontier technologies. Several new technologies arouse anxieties of various kinds in the minds of many because of a difficult to predict trajectory of their growth and consequent implications — particularly technological products replacing jobs which human beings can do. Most of these threats are directly related to high technology, besides of course some social factors. Technologies related to climate control, artificial intelligence, internet of things and additive manufacturing may have a drastic impact on the livelihoods and lifestyles of millions. Various nations are affected differently by these events, but a truly global nature of their impact may lead to a coordinated response akin to what was observed in the case of the Paris Accord on climate change. In Paris, a large number of countries fearing global impact of climate change agreed to initiate steps to bring carbon emissions within permissible limits. In the case of climate change, it was easier to appreciate the link between local, regional and global impact and convince the world community to act in unison. Even in that case some discordant voices have been rising. Organising a collective response to the challenge of several new disruptive technologies may not be that easy. Developed countries sense an opportunity in the development of many of these technologies related to artificial intelligence and robotics, pressured as they are with large scale offshoring of manufacturing to many Asian nations. Some of them are also unhappy about the decline in the quality of products thus built. Many of them are working towards developing onshore manufacturing facilities based on these technologies. They also feel the heat of probable competition from the large additional labour force from countries like China and India in the near future, particularly as these nations are reasonably advanced in terms of technology development. Many issues of common global interest in deployment of these technologies also relate to ethics, security and regulatory concerns. For developing countries, these technologies could also threaten the employment of their labour force. These concerns cannot be brushed aside as they may cause social unrest and trigger illegal migration across borders. The divergence of responses of developing and developed countries to these developments is quite probable.

A major problem in developing countries is the switchover of employment from agriculture to manufacturing and services. The labour force released from agriculture may not be qualified to handle new jobs in manufacturing and service sectors without appropriate training. In the case of sectors using radically different technologies, an overwhelmingly large workforce would require training. Organisation for Economic Cooperation and Development OECD) reports indicate that in most developed countries, there has been a steady and persistent decline in the fraction of labour force employed in the manufacturing sector – from 30-35% about three decades ago to about 10% presently. However, in developed countries, jobs have reduced in the manufacturing sector primarily because of strategies of multinational corporations (MNCs) that have tried to shift operations to low wage countries. While traditional goods manufacturing has suffered a decline, high technology products have grown in their contribution to generating employment. Developing countries may inevitably face a similar situation with some time lag – fading away of low technology/skill jobs. It is evident that most of these disruptions or threats can be countered by banking on increased scientific research and development (R&D) and encouraging innovation and wisdom in deployment of new technologies. Disruptions have to be understood in their local context at the level of science and technology for the sake of appropriate response.

An important question in this context is what kind of scientific research and technology development needs to be encouraged. Developed countries seek "cutting edge research" and innovation in producing novel products or more efficient production means. This helps them in staying ahead of competing countries and ensuring markets for their own products. Many developed countries are enhancing their investments in what is called "knowledge based capital" (KBC). This type of capital is different from physical infrastructure and is related to software, database, upskilling of workers, market research, financial services, copyrights, designs etc. These are extended arms of scientific research today. The answer to these challenges is more innovation, science and technology and not less. Developed countries have realised that KBC together with scientific research is the key to prosperity. This realisation is yet to dawn upon most developing countries. Most developing countries do not have sufficient funds and cannot afford so much investment on R&D and education. These countries are influenced by several international trends and they often fall prey to imitating developed countries in pursuing similar kinds of scientific research. This may not be the best recipe for them and usually it turns out to be unequal competition and "wasteful research." The real advantage of research can be exploited only by making early and speedy translation of results into technology development and deployment. Many developing countries give undue emphasis to frontier R&D which is of doubtful value to them. In the process, they are marginally able to attain the status of the "also ran" category at the cost of their precious resources. There is another set of countries which stress heavily on defence or strategic R&D, thus spreading their resources thinly and neglecting industrial R&D. It has been observed that it is often more useful for such countries to spread technology and its benefits, not necessarily high technology, horizontally to wider populace. A wide range of dispersion in deployment of technologies in these countries because of variable technology assimilation capabilities of different entities causes huge loss of potential benefit. The inability of nations to widely spread benefits of technology due to various reasons such as lack of trained human resources or assimilation potential needs to be addressed. Maintaining a balance between basic and applied research, between technology development and deployment and establishing appropriate linkages between research and manufacturing is the key to their progress. This fact needs to be ingrained in the institutions of these countries. There are a variety of inhibitors which restrict science-linked industrial growth potential of developing nations.

What are the systemic inhibitors that do not allow many of the developing countries to harness their full potential for benefit of their own population and global ecosystem of innovation and manufacturing? These answers cannot come from standardised indicators and rankings. Many of the macro indicators in developed countries can at best serve as broad guidelines for the policymakers of developing countries. This is particularly true for areas of scientific research, technology development and Science, Technology, Engineering & Math Higher Education (STEM HE). Unravelling and overcoming these inhibitors requires understanding the anatomy of institutions and systems, dynamics of processes and cultural legacies. Even within a country, different institutes could fare differently on these yardsticks.

Most of the developed countries have several common features like better higher education, active STEM R&D institutions and private industries involved in research to carry out innovations in products and processes. Governments in these countries help such industries in expanding and eventually becoming MNCs. These industries are driven by profit and invest heavily on innovation and R&D and own proprietary rights of most new technologies. Most of the developed countries are well knit with the help of Organisation for Economic Cooperation and Development (OECD) – an organisational structure which has fertile R&D

groups working to guide them on policies and actions. OECD has comprehensive databases and prepares analytical reports about economy, education, STEM R&D, technology and its impact on development. Developed countries also share the best practices between themselves for maximising returns on investment. Over the years, they have developed several broad commonalities as far as their economic and scientific culture is concerned. They have managed to do that despite their geographical distance, difference in culture and level of growth.

Developing countries have their economy dependent mostly on commodities or low and medium technology goods and services. Some of these countries are progressing relatively faster in terms of technology and related manufacturing and services. They have a nascent grouping called BRICS, comprising Brazil, Russia, India, China and South Africa, on the lines of OECD, but with no comparable success so far. The interdependence of the two groups of countries manifests itself in a variety of ways. Demand for high technology products in developing countries sustains employment in developed countries producing them — this may include products ranging from defence technology, consumer items to machinery for producing medium technology goods and services. While developed countries set the trend for technology development related to these products, users in developing countries simply follow these trends with a time lag depending upon their level of growth. Many advanced technologies like robotics, artificial intelligence, internet of things, autonomous vehicles etc. are being developed with an aim of ushering in a revolutionary change in manufacturing, services and even lifestyles. These technologies, depending upon the trajectory they take, may have the potential to harm the interests of developing nations as the developed ones are likely to have an alternative to cheap manual labour and will be less dependent on developing countries.

Developing or less developed countries can always learn from their developed counterparts. Many a time, the lesson is to avoid a particular trajectory of growth. It is like an opportunity to leapfrog by bypassing relatively obsolete technologies — taking advantage of fast obsolescence of many modern technologies. Many developing countries may be confronted with common situations even in the choice of R&D and the areas of priority. There is need for innovation and for the science & technology (S&T) community in these countries to broaden their horizon and to cover novel and original R&D problems rather than only well acknowledged international frontline problems. These problems could be rooted in local situations or require distinctive deployment strategies. Solutions to these problems may trigger a more inclusive and thus sustainable growth. Some of these could have multiplier effects in other developing societies too. It would require a mindset change for researchers, decision makers, and funding agencies to support such research. Originality in choice of research must get its due respect vis-a-vis international frontline.

Several BRICS countries can lead such scientific research and technology development, and act as role models for other developing countries. The basis could be commonality in problem areas, proximity of development status (or level of poverty) and similarities in governance structure. Most of the BRICS nations are interesting candidates for studying R&D and manufacturing ecosystem and how they attempt to compete with or mimic the ecosystem of developed countries.

Among BRICS nations, China and India together constitute about 36% of the world's population and are progressing rapidly in terms of scientific research and technology development. China has gone far ahead and soon may be counted among the developed nations and has a unique, but monolithic governance model, impacting growth favourably. India closely approximates as a nation with which a large number of developing countries can identify themselves in many of these respects. India with its broadbased R&D, infrastructure for STEM HE and level of technology development is in a position to act as a role model for many developing countries in encountering the challenge of disruptive technologies. India is a very heterogeneous country covering several small 'sub-nations' (various Indian states) within, with widely varying levels of economic and technology development. Presently, India is at a crossroads as is gradually shifting from agriculture to manufacturing and services. While jobs in manufacturing are growing at a slow pace, service sector's contribution to the Gross Domestic Product (GDP) is growing much faster. Hi-tech manufacturing is not able to grow fast. Both service sector and hi-tech manufacturing require qualified workforce and it seems that presently, the service sector is sucking qualified workforce at the cost of hi-tech manufacturing.

India along with many other developing countries, while aiming to develop science and technology-based manufacturing ecosystem, needs to be conscious about some social conditions/constraints. The objectives have to be consistent with the needs of poor and dispossessed citizens, especially those living in remote areas — people who are exposed to raw elements of nature, but are not well connected to normal supply chains. Secondly, since governments of many these countries cannot afford to spend heavily on R&D, higher education and knowledge-based assets, these resources have to be used judiciously and there should be a collaboration with private enterprises for investment in these areas. It is quite often that while private enterprises reward innovators, they also have to take care of their investors, who may demand a level of profits not affordable to the poor consumers/users of these products of innovation. It is the responsibility of these governments to strike an optimum balance between encouraging innovators and investors on one hand and ensuring fair cost to consumers on the other.

India has developed a few islands of excellence in STEM HE and technology hubs with a large fraction of population divorced from its benefits. It has a large number of STEM HE institutes but only a few of them are world-class. It has some robust R&D institutions working in strategic sectors like atomic energy and space, but a much less industrial R&D with meagre contributions from private sector. There are many universities involved in R&D with ad hoc choice of R&D directions and poor linkages to industries. A sort of myopia and incoherence prevails, which while pointing at certain shortcomings, indicates big potential from the overall ecosystem, if reformed suitably. The existing STEM HE system, type of R&D directions and technologies being developed, lead to a poor match of supply of skills and knowledge base and their demand among manufacturers. This often results in scarce talent moving out of

India. This discourages promotion of favourable ecosystem and leads to decreasing interest in the younger generation in these subjects or even working within the country. Such migration of talent from India and other developing countries to the developed world should be a serious concern. Ross Basset in his book *Technological Indian* has recorded the history and progress in careers of a large number of Indians who went to MIT, USA almost a century ago. . He has described in detail how many of them impacted the global technology scene significantly. But these talented Indians could not be harnessed within India. Similarly, in more recent times, a large number of migrants from Indian Institutes of Technology (and many others) have impacted the Silicon Valley's startup ecosystem. These observations indicate the type of potential that exists in India for growth in STEM HE, scientific research, technology development and manufacturing. It also points out that if India and other developing countries are able to develop the right ecosystem to attract and retain talent in face of competition from developed countries, then that may impact science and technology-linked manufacturing growth in these countries. This may influence the global distribution of manufacturing of hi-tech products.

This work aims to understand the anatomy of scientific research — STEM HE, and the manufacturing ecosystem in India. There is need to transcend the simplistic indicators of gross progress in scientific research and manufacturing. This may involve deciphering dynamics of institutions and processes, understanding the culture and legacies, policy structures, accountability mechanisms and related actions (or inactions) and the relevant socio-political milieu.

The second chapter of this work describes how scientific research and technology development has influenced the manufacturing scenario globally and how low and medium technology manufacturing has shifted to some of the Asian countries in recent times. It also points out how the rapidly changing global technology scenario is both a challenge as well as an opportunity. How different countries are responding to this. It also points out how high technologies and startups based on them are bigger contributors to employment. The challenge becomes more daunting for developing countries with large population/workforce which is not suitably geared up by way of knowledge base and skillset. It describes how the capability to attract talent from all over has contributed to technology development in developed countries. This suggests how some of the countries like India can act as a bridge between varying levels of technology growth in developed and developing countries. India itself is a major potential growth story in manufacturing with as yet poor growth of its STEM talent. This offers itself as an interesting case study.

Chapter 3 briefly describes the linkage between scientific research and manufacturing in India. The Government of India is promoting manufacturing through its mission 'Make in India,' with one of the major objectives being jobs for millions. The role of scientific research and technology development is often not well appreciated in achieving this objective. This chapter describes in brief the spectrum of manufacturing sector where R&D can play a worthwhile role. This chapter also attempts to bring out some examples of futuristic as well as some neglected themes that link science, technology and manufacturing.

Chapter 4 describes how some of the major Indian R&D labs and higher education institutions have evolved, their present structure and mandate. It shows how India has been able to develop its infrastructure for scientific research after independence. How different organisations are funded and supported. It also describes how this infrastructure is not yet harnessed well enough for promoting industrial growth. This chapter aims to explain how it may require many of these labs to reorient themselves and shed off some of their cultural and legacy limitations for the sake of new goals.

The fifth chapter deals with issues related to the limited involvement of large sections of private industries in R&D and also their poor linkages with government R&D labs. It points out how a substantial part of the Indian industry belongs to the small and medium sector, which is very low in productivity because of poor or sometimes no background in innovation and R&D. The chapter explains the need for evolving an efficient symbiotic arrangement between various players involved in R&D and manufacturing. These entities could be big or small, private or public, young entrepreneurs or well established players. Each of them has a role to play and needs facilitation. It describes how a situation can be orchestrated, using competition as the key, to stimulate private sector R&D.

The sixth chapter talks about the policies that influence scientific research, STEM HE, technology development and the manufacturing sector. It throws light on how all these can be brought together to ensure optimum output generation. It points out how missing even one of these entities can derail the overall process of high technology-based manufacturing. It talks about bringing STEM HE on par with other national strategic objectives. There is need to bring some policies on commerce of hi-tech products in sync with indigenous technology development. Many of these issues require a holistic vision and therefore it is required to have a high-powered National Science Technology and Innovation System. This body needs to go beyond playing an advisory role and subsume multiplicity of authorities that drag down both policy making and accountability. A few suggested roles and functions of this proposed body are described in the chapter.

India and many other developing countries that became independent around the middle of the last century or even later had neither participated in nor witnessed the industrial revolution. This has not only stunted their scientific and technology growth but has also created a cultural void among their industries, scientific organisations and citizens in general. In some countries that have struggled and made progress, some of these gaps have been filled. The seventh chapter discusses the cultural aspects of what is often called nebulously the ecosystem. There are wide ranging issues that need attention for the healthy growth of science, technology, and manufacturing. It could be bureaucracy of all kinds inhibiting innovation-led growth, loss of STEM talent to developed countries, undue respect for foreign technologies and degrees, inability of STEM HE system to cope with issues of quality and quantity, unfair performance appraisals of projects and individuals and many such other issues. Many of these issues have led to limited confidence in policymakers in Indian R&D establishments despite some of their outstanding achievements.

The eighth and the last chapter attempts to summarise and describe some of the salient points of discussion in earlier chapters, mentioning possibilities of generating accountable mechanisms to evolve a genuinely symbiotic ecosystem of STEM HE, scientific research and technology development in both public and private sectors as well as manufacturing industries. It points out how despite several gaps, the potential for growth is huge. Many of these issues are common in developing countries. India could offer an opportunity to such countries to join hands and work together to face challenges of development in face of disruptive technologies.

Chapter 2

SCIENTIFIC RESEARCH AND MANUFACTURING: GLOBAL CONTEXT

INTRODUCTION

A major part of the twentieth century witnessed industrial manufacturing growing fast on the back of scientific research and technology development. This had a strong impact on economic well-being and employment across several nations. While scientific research and technology development are fountainheads of industrial progress in general, it is the non-science policymakers who are mainly responsible for influencing major related policy decisions. Some of the nations that have given due weightage or active support to scientific research as well as proactive participation of technocrats in relevant policy making, have made rapid strides in manufacturing. A few good examples are Japan, Germany, Korea and Israel, where strong institutional mechanisms exist to ensure this and they spend a fair fraction of their GDP on research and development. Presently, there is a stronger need to involve scientists and technologists in futuristic planning as global manufacturing may be about to usher in a revolution because of the advent of several new and potentially disruptive technologies. This may impact lives of millions of workers and lifestyles of consumers, producers and users of novel products. The nature of impact cannot be predicted with certainty, but governments will need to be alert to the possibilities, trends and pitfalls.

Towards the end of the last century, a steady but persistent decline was observed in the manufacturing sector in developed countries. In the seventies of the last century, employment contribution of manufacturing in these countries ranged from 25 to 35%, which declined to about 10% or even lower in some cases [1]. In contrast, major developing countries like Korea and China witnessed steep growth in manufacturing employment percentages. While the rise of Korean manufacturing started in the early eighties, it peaked by the early nineties and thenceforth followed a slow declining trend. In the case of China, growth in manufacturing-based employment started in 2000, and has reached more than 30% of its (GDP) [1]. Many other Asian countries have made good progress in manufacturing like China and Korea, but more on the strength of low wages of their workforce, improved infrastructure and regulatory regime and less on the basis of genuine technology growth. The general trend indicates a shift in manufacturing from developed countries to low-wage developing countries. A more significant trend noticed in most developed nations is that while the overall manufacturing employment share declined, it increased in the case of R&D intensive industries [1]. These are, for example, those related to chemicals, pharma products, aerospace, computer etc. On the other hand, traditional products like textiles, metals, and plastics have fallen behind in their share of employment. There has been a constantly rising trend of R&D-linked manufacturing in Korea and China. Korea, which is among the biggest spenders on R&D (as a fraction of GDP), has seen continuing growth of R&D-based industrial employment. These observations reflect the fact that the share of developed countries in global exports of such products has fallen from 73% to 55% from 2000 to 2013 [1]. Figure 1 shows the trend regarding R&D-linked manufacturing-based employment for many developed countries from the eighties to 2013 [1].

It can be seen that among the developed countries, Germany, United Kingdom and the United States have more or less retained their share at around 30%. It may be difficult to pinpoint specific reasons, but there are a few distinctive features of each of these developed countries which suggest how they have been able to maintain their level of contribution to R&D-based manufacturing. For example, USA has been pioneering in industrial research labs and attracting high potential talent from across the world. It has a well-developed venture capital system to support innovation activity. In the case of Germany, it has focused to encourage innovation in small and medium enterprises (SMEs) and smooth transfer of R&D results and technology from labs to these units through institutions like the Fraunhofer Society. Such growth takes care of a large fraction of workforce enabling innovation and inclusivity to be consistent to each other. Similarly, Japan takes extreme care through its Ministry of Technology to involve scientists, technologists and industrialists to plan the national strategy and develop appropriate human resources. It is obvious that each country has its own distinctive strong points in harnessing science and technology, enabling it to grow. High technology is evolving rapidly and every country needs to develop its own response to the challenge.

Figure 1. R&D intensive manufacturing-based employment as a fraction of the total employment in a few selected economies from 1980 till 2013.

Managing Global Technology Challenges and Scientific Research

The rise of high technology and knowledge-based industrial manufacturing and its various facets could be treated as a challenge or opportunity depending upon how one looks at it. With an aim to exploit high technology-based manufacturing and remain competitive, most of the developed countries are investing heavily in scientific research and technology development. The Organisation for Economic Cooperation and Development (OECD) countries have fixed a target to spend 3% of their GDP for research purposes. Most of these expenditures have more than 60% contribution from business sectors and the rest is divided roughly equally between government research institutes and educational institutions. Such has been the nature of modern technologies that investment in basic sciences is progressing faster with respect to applied sciences which is counter intuitive and is due to anticipation of novel breakthroughs from such research. Countries like China are fast closing the gap on this front with developed countries. Despite investment in science and technology, translating research results for promoting manufacturing, job creation or public good may not always be equally effective in different countries. Some of these countries have harnessed their scientific research and technology development efforts more efficiently for the sake of industrial progress. This efficiency is dependent on actual performance as well as salaries of scientific workforce. Average emoluments of scientists vary widely from one country to another and constitute a major fraction of investment. Success in manufacturing novel products is dependent on harnessing R&D results suitably. In the present day manufacturing, while patents are primarily responsible for novelty of technology, there are many other aspects of knowledge base which contribute significantly to the maturing of the final product. This may include associated proprietary software, trademarks, designs, human resources, organisational capital etc. These are labelled by economists as "intangible assets" or also as "knowledge based capital" (KBC). Businesses and countries are gradually realising the significance of such KBC in improving competitiveness, innovation and productivity. For example, USA and UK invested 1.5 times more in such capital than in physical assets in the year 2013. France and Germany made about 80 to 90% of their investment in physical assets in that year and this trend of investment is gaining acceptance and growing rapidly [2]. Another prominent trend is that of young firms, born less than five years ago, being good repositories of such knowledge base. It has been found that these firms are much bigger net creators of jobs as compared to the older (well established) and much bigger companies. In a study carried out in OECD countries, it was found that such startups were responsible for on an average about 18% of total employment in the country, but contributed to net creation of 47% of new jobs [2]. Construction, manufacturing and service sectors were covered in the survey. Figure 2 shows such behaviour in a few chosen developed countries [2]. This strengthens the impression that well directed R&D and KBC can be big contributors in creating new jobs and thus countries need to increase investment in these fields.

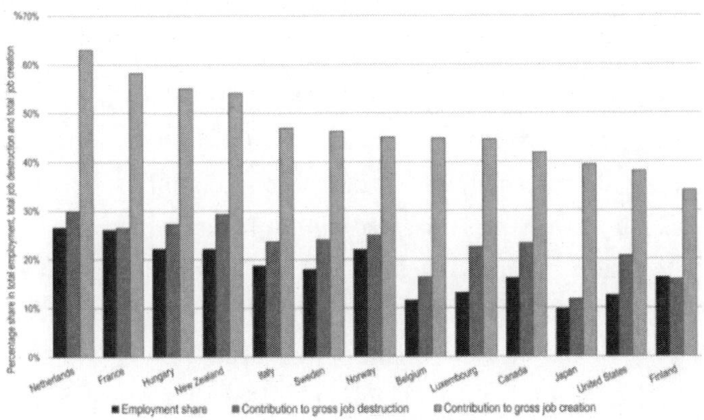

Figure 2. Contributions of young technology-based startup companies in job creation, their relative role in overall employment in a few selected countries.

As can be seen there is a strong linkage between R&D, innovation intensive manufacturing industries and job generation – contrary to the impressions in many developing countries struggling to shift their economies from agriculture dominance to manufacturing and services. These impressions are based on poor experiences of “inappropriate” growth of R&D in manufacturing ecosystem. The countries that have reaped dividends from investment in the R&D system and KBC continue to strengthen it further. Business organisations in these countries are at the forefront of this process. This difference in emphasis on R&D between countries is one of the major reasons for growing global technology and economic disparity between nations. In fact, presently, there are some sorts of monopolies of technology worldwide. About 250 MNCs contribute 70% of total R&D expenditure, they own 70% of patents, 80% of Information & Communication Technologies (ICT)-related patents and more than 40% of trademarks. Most of them have headquarters in USA and Japan. Two thousand top R&D spenders own 90% of global patents and trademarks [1]. Most of these are private entities and they contribute predominantly to value/employment generation in their home countries. For example, Apple's iPhone 4, which was available at a retail price of 600 dollars, yielded 270 dollars to the company. The Korean companies supplying its core components received 80 dollars and the Chinese companies which assembled the phone got about 1%. But these companies work with utmost loyalty to their investors. In the garb of protecting innovators, they tend to benefit their shareholders too. Sometimes this could clash with the interests of poor consumers. This is noticeable in the case of providing access of life-saving drugs to poor patients in developing countries. There could be many more such examples. Thus, evolving a balance between interests of innovators, consumers and investors is a challenge in the long run. Reducing monopoly in any technology area will be good for inclusive and sustainable growth and the betterment of mankind.

Globalisation of markets has been a challenge for many countries, but an opportunity for some others. In economic terms, it has influenced demand for novelty/high technology products and consequently the job scenario. In terms of technology, it has spurred the production of high-tech gadgets which have become trendy. Demand for hi-tech products in more populous developing countries has generated jobs in developed countries producing these products, outsourcing of low technology products to low-wage countries generated jobs in developing countries. But the number of jobs required in developing countries is much larger. Consequently, poor economic conditions and lack of employment opportunities have upset many developing economies triggering migration. In the process, the talent required by developing countries is also lost, compounding the problems further.

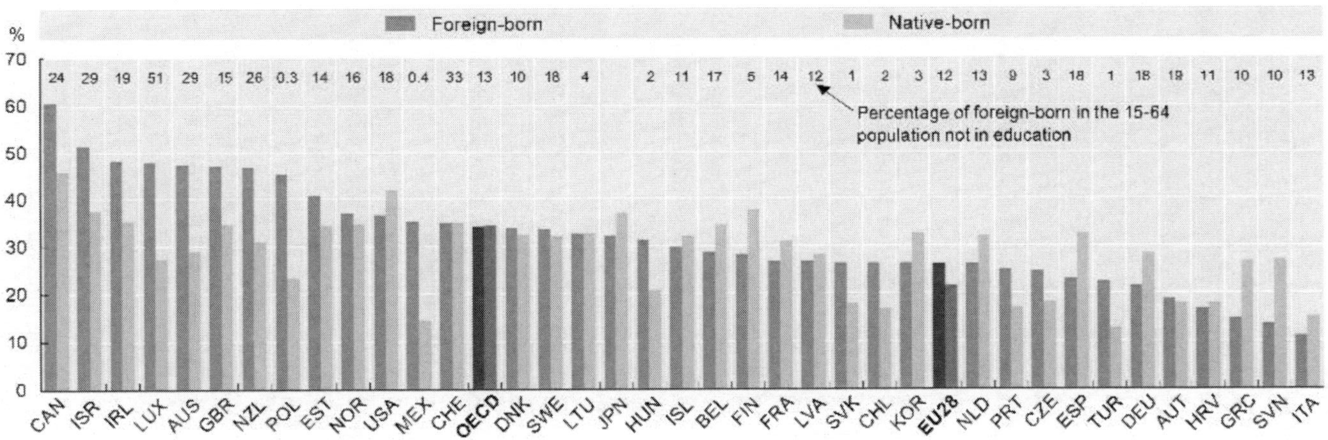

Figure 3. Highly educated individuals in immigrant and native population in few selected countries (acronyms explained separately).

Besides the economic demand scenario, job situation and talent migration, the nature of scientific research is also affected by the globalisation process. Scientific research in developed countries influences the course of R&D directions in developing countries too. While basic science is universal in nature, technology developments need not always be so. These globalised influences tend to divert attention of scientists and technologists in developing countries from the problems rooted in their own soil, which do not happen to be trendy internationally. International influences encourage growth of what can be called aping of frontline research by developing countries. Such research is related purely to publications and related citations. Such research may have marginal impact on the growth of developing countries. For example, poor progress in developing cure for tuberculosis, malaria, dengue, malnutrition etc. Many Asian and African countries bear testimony to this neglect. Ironically, the advent of solar power happened in relatively sun-deficit developed countries rather than in sun rich equatorial countries. Internet connectivity research being done by MNCs in developed countries for remote areas is aimed to tap markets in developing countries. This could have been initiated in less developed countries (better developed among them), but that did not happen. Sometimes the reasons are related to shortage of available funding but more often it is otherwise. Another serious aspect of the problem is that in the absence of a robust database of their own, these countries tend to rely heavily on international databases for information such as rankings of educational institutions or usage of scientific research and technology which are used as inputs for policy making and actions. The main objective of such rankings is to offer a suggestion to policymakers. Many a time, the analysis as well as emphasis given to various factors is applicable more to developed countries than poor and developing countries. Policymakers in developing countries are likely to be misled in the process. This could impact their trust on the genuine domestic demands for R&D. Talent attraction and retention, particularly of scientists and technologists, has been one of the pillars of success of many developed countries. In some of these countries, science and engineering doctorates of foreign origin outnumber the natives. Figure 3 shows the ratio of highly educated migrants and natives in some of the developed countries [1].

This policy of attracting foreign talent is normally pursued despite protests from the locals and loss to the source countries. In recent times, some of the developed countries have started reversing this policy with an aim to promote interests of domestic talent. Such policy measures might influence migration of talent between countries. After framing of certain laws in England about foreign students not being allowed to work after completion of their course, there has been a steep decline in students from South Asia applying for admissions into British universities. Similarly, a number of talented migrants to USA have been trying to return to countries like China and India, after certain new policy measures against migrants were adopted. A decline of 7% in admission of international students to American universities was reported in 2017 [3]. Protectionism of this kind could indirectly benefit developing countries, provided they can leverage returning talent suitably. Their capability to attract talent is crucial as otherwise there are more countries like Canada and Australia waiting to tap them. This is only a small part of talent migration problem and if developing countries wish to retain their own scientific talent nurtured with public support, then they must improve their domestic ecosystem of scientific research, industrial manufacturing and suitable employment. Developing a balance between free international movement of talent and retaining talent for attending to indigenous R&D and manufacturing in source countries is a challenge which may require global attention, besides that of the affected nations.

A challenge (perhaps opportunity) of a different kind may be taking shape for the global science, technology and manufacturing ecosystem. Several new technologies are emerging on the scene which are influencing (or likely to do so) the way we manufacture and the way we work and live. These technologies have the power to disrupt even the employment scene in a major way. These technologies are broadly of two types. Firstly, there are technologies which are novel and are required urgently and have the potential to create new types of jobs without destroying the existing ones. Environment monitoring and control on regional, national and global levels is one such task. For this purpose, a variety of tools and technologies are being developed worldwide. Similarly, cyclic economy, which aims to retrieve value from wastes of all kinds, is another example. In both these cases, new jobs are likely to be created without destroying the existing ones. There could be many such examples where jobs are mostly based on new knowledge base. These are fit cases for developing countries to emphasise upon. Secondly, there are technologies based on computers or internet, like for example artificial intelligence, robotics, internet of things, auto-vehicles etc. These are the kinds of technologies that are likely to destroy some jobs like driving, low-level surgeries, simple machining, and a variety of office and household tasks. But they may also create totally new types of jobs. While new jobs are expected to require higher level of skills and knowledge base, jobs being destroyed will be of lower level skills. For developing countries, this is a major challenge. Unless they invest in generating appropriate knowledge base and skills, they could be big losers. For countries like India that have a reasonable background in information technology, this could be an opportunity if several aspects of research and technology development are handled with imagination.

RESPONSE OF THE NATIONS

How different nations are preparing themselves to face these challenges and disruptions. Besides technology there are some disruptions related to social affairs impacting scientific research. Large scale migrations following social unrest or wars, ageing populations in technologically advanced nations and unemployment in populous countries are some of the factors giving additional push to technological disruptions and talent migration. Every nation perceives these challenges differently, depending upon its internal situations and legacy. One can sense some kind of polarisation happening on some of these issues over a period of time between rich and poor countries. This may be unlike the challenge of probable climate disruption as scientific studies established the threat impacting most of the nations — some more and some less. This could result in Paris Climate Deal with each country willingly compromising on its original position.

One must analyse the origin of divergence of perceptions about these challenges in greater detail. In this context, it may be interesting to recall what economists call the "North South Trade Model" [4]. This model postulates that developed countries or advanced area (North) has the skilled workforce and R&D capability to innovate new goods and services, while the less developed countries (South) cannot compete in these areas. As a result, North innovates new goods and trades them with South, which produces older goods as it gains the technology to do so. Once the two regions have access to the same technology, the wage advantage to workers in North will be lost. Workers are paid higher in North than in South both because they are more skilled and because North has a monopoly on new technology products. More rapid technological advance increases wages in North as compared to wages in South, while more rapid diffusion of technology has the opposite effect. If this is the case, developed countries will seek innovation in new upcoming technologies but may not like to part with the latest aspects. On the other hand, developing countries will continue to seek transfer of these technologies, besides efforts to develop the same.

With this background, how are some of the major countries of the world poised to handle development of new technologies and thus raise the standard of living of their workforce. How much do they spend on R&D and innovation and how many workers are involved in these activities – as these are major inputs for development of technologies. Different countries invest differentially into scientific research and technology development and inputs vary drastically and are predominantly related to research funding and number of researchers per thousand of their population. For example, India spent 0.69% of its GDP on R&D in 2014-15, Israel and Korea spent more than 3%, most developed countries spent above 2%, Russian Federation 1.19%, China 2.05% and so on [5]. Figure 4 shows the spread of R&D expenditure/GDP for some selected countries [5]. There is correspondence in scientific and technical manpower per million of population. While countries like India have about 218 researchers per million of its population, Israel had more than 8000 per million of its population [5]. Most developed countries have several thousands of scientific researchers per million of population. These figures need to be considered along with the salaries of scientific workforce. Developing countries have very low salaries — the ratio is one fourth to one sixth between American and Indian researchers.

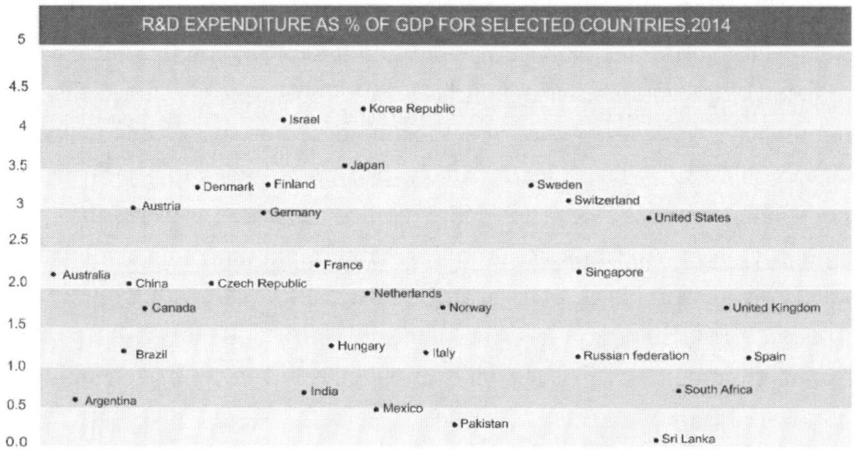

Figure 4. Expenditure on R&D as fraction of GDP for a few chosen countries — both developing and developed.

What is the outcome of these investments in the form of publications? Besides publications, citations are also very important. But open access literature can only be broad indicator of R&D for the sake of manufacturing and overall it is much more complex to estimate R&D, innovation and their linkages to manufacturing in individual countries. For the sake of manufacturing, patents are considered better indicator than open access publications. Most of the new patents and trademarks are presently owned by USA and China, followed by Germany, Japan and France. A more significant parameter of interest is how much different countries are able to advance in the field of new and disruptive technologies. The results are no different and the same set of countries continues to dominate.

This pattern confirms the fact that rich countries will usher rest of the world into a new and perhaps disruptive technology regime in the near future and poorer countries will either lag further or maintain the gap of technological progress. Of greater concern is the potential impact of this on the employment scenario. This will necessitate developing countries to find ways so that they do not continue chasing the mirage of following the footprints of research in the developed world. While diffusion of technology products cannot be restricted because of free trade requirements, diffusion of technology itself is restricted because of patents and trademarks. Developing countries may explore the route of collaboration and foreign direct investment (FDI), instead of competition, but cannot be assured of reaching the same level of technology. Is it possible that these countries begin exploring new avenues of innovation and R&D related to their localised problems which are not addressed by the present-day international R&D ecosystem, instead of continuing on the path of collaboration in traditionally acknowledged research areas. Such efforts may call for path breaking originality but they do have the potential to be rewarding.

Why are these countries not able to make progress independently on scientific research and impact manufacturing? Each country has its own set of reasons. Among developing countries there is large variation in the level of progress in scientific research, manufacturing and their linkage. Leading the pack are Brazil, Russia, India, China and South Africa (BRICS). These nations have made major strides in science, technology and hi-tech manufacturing. China starting from producing low technology goods has fast matured towards high technology products and is close to being labelled a developed country and in some sectors hopes to pose a challenge to USA. Each of these countries has excelled in few isolated areas and do face a multitude of problems in developing a genuine all-round science-technology hi-tech industrial manufacturing ecosystem. Russia, for example, has well developed foundations of basic science, but has a legacy of heavy stress on military and space technologies. The inability to translate scientific research into industrial manufacturing and profits has been a bane for them. There is a famous example regarding Nobel Prize for invention of laser. This prize was shared by three scientists, two of them Russians — Basov and Pokhrovich, while the third one was an American, Charles Townes. It was Townes who decided to patent his invention, while Russian scientists never thought of it. Today, laser-based industries run into billions of dollars and Russia's loss is obvious. There are several such cultural traits which afflict some of these nations. Besides culture, dynamics of processes, chains of legacy, choice of R&D directions and governance of R&D and industries and many factors related to economics and regulatory structures have their own role to play.

There are many other developing nations besides BRICS countries which face these challenges. One can distinguish between developed and developing countries by noticing their gross expenditure on R&D as a fraction of GDP. As Figure 4 shows, most top rows are occupied by rapidly progressing countries. Most of them have manufacturing based on high technology. Another criterion is gross enrolment of students in STEM HE [6]. A clear correlation between STEM higher education, gross expenditure on R&D (GERD) and status of economic development is noticeable.

INDIA AS A ROLE MODEL FOR DEVELOPING COUNTRIES

In their efforts to build a hi-tech R&D hi-techy-tech industry ecosystem, some of the BRICS nations can set examples for the less developed nations. India, among BRICS nations, can be a good role model for less developed countries in building R&D industry ecosystem. This is while India itself is struggling to come out of the legacies of a variety of chains pointing out to commonalities of challenges with many other developing countries. One significant characteristic in case of India is that it can act as a bridge between developing and developed nations. An example of this is how Indian IT industry saved the situation of Y2K bug which had a significant impact on the developed world. Its IT industry has been serving developed nations for quite some time in a mutually beneficial model. Similarly, its pharma industry has been able to develop cheap generic medicines which impacts large poor populations in developing nations. India has been labelled by some as "poor man's pharmacy." This characteristic makes it a partner for less developed countries in the joint struggle and a bridge between developed and developing countries when it comes to high technology.

INDIA: AN UNFOLDING GROWTH STORY

How India as a nation presently perceives innovation, scientific research and technology development as well as its role in linking hi-tech manufacturing and inclusive and sustainable development. It may seem that more often than not, the Indian political class is usually clueless and the supportive bureaucratic structure is ignorant (or misinformed) in a large measure and form principal constituent of policy making structure. More significantly, they are negligent about performance outcome of R&D projects and focus primarily on financial closure. Failure in major R&D projects is accepted as a norm without earnest questions. They believe that to support scientific R&D, a predominant input is funding. This is not often realised that appropriately matching STEM higher education and skills to demand of R&D institutions and manufacturing industries is crucial. Even if they do so, coordinated action is wanting, as a holistic vision is rare. It is not often appreciated that any sectoral R&D cannot be made available off the shelf. Some of these fallacies and myths about scientific R&D in policymakers can impact negatively the growth of R&D-manufacturing ecosystem. All advisory bodies do have limitations as per their limited mandate. Indian S&T community leadership usually falls in line and is not very proactive and forthcoming to guide policymakers for the sake of building a hi-tech industrial base. There are no systemic mechanisms in place. These institutions do not consider it their responsibility as they work within sharply mandated silos, advisory councils notwithstanding. This influences to a great extent ground level researchers and their R&D directions. Indian R&D-based manufacturing has many distinctive features of its own, both good and bad. Its R&D investment is heavily biased in favor of government expenditure. An indifference of decision makers to R&D as a means of progress in social goals is indicated by the fact that for a long time, GERD has been static at about 0.8%, irrespective of the political colour of changing establishments. This is some sort of vote of limited confidence on Indian researchers. A further weak link in the growth story is that the government' effort suffers from myopia of strategic technologies somewhat akin to that of Russia. In the process it has made good progress in space and nuclear technology but at the cost of poor progress on technologies for industrial manufacturing. Despite this emphasis, defense technologies, which are specialised class, have not progressed as expected. While the former two sectors have relative liberty to progress at their own pace, defense demands are quite different in nature. All government R&D institutions garnering major funding collectively have a perhaps unacknowledged responsibility. This is to enhance the trust of public and also industry in the worth of innovation and scientific R&D. This is particularly true for countries like India that still have a poorly developed innovation ecosystem. This poor level of confidence in the potential of R&D to enhance profitability is at the core of the industry's lack of willingness to invest in R&D. Developing a healthy ratio of R&D expenditure between public, business sectors and STEM HE institutions in a fruitful and symbiotic manner is a target worth pursuing. A holistic review of R&D, STEM HE and hi-tech manufacturing in India and many other developing nations can be quite fruitful for their progress, bridging the divide with presently advanced nations and bringing forth distinctive novel technologies or even their mode of deployment.

India has a good stock of talent which is not harnessed well within its boundaries. But in the process, it has indirectly contributed to building a fertile diaspora. India is a vast nation with several heterogeneity and complexities within. The society is deeply attached to traditions. This impacts even the segments related to science, technology and industry. For example, as a part of its deep-rooted "Brahmanical tradition" (elitist in Indian parlance), there is heightened respect for basic and conceptual aspects of sciences as compared to instrumentation or product development and related commerce. Vocational education is valued much less compared to say graduation in arts. This, despite the fact that the average market value of graduates in arts is much lesser than that of those with diploma from vocational institutes. Indian contribution to basic science publications is rising at a fair rate, while its hi-tech manufacturing is declining. There are several other factors at work. Respect for foreign technology is irrationally high and this discourages demand for indigenous technology, leading to a vicious trap. There is excessive emphasis on traditional manufacturing sectors like textiles and leather, assuming these are employment extensive. While evidently these sectors have grown in automation, (not realising that automation is mostly based on imported machinery!) and there are other neglected sectors with higher job potential, this realisation is yet to dawn upon many. India is thus finding it difficult to graduate from low or medium to high technology sectors, partly because of lack of skilled workforce. Pertinent STEM HE and related R&D which are expected to be fountainheads of hi-tech manufacturing are ignored by its universities. About 80% of funding is absorbed by few central universities with only 5% enrolment of students [6]. India has the basic education infrastructure and is among the countries with the largest number of STEM HE institutions in the world, but these institutions suffer from acute lack of quality faculty. Vocational education is neglected and not valued socially — with blue collar jobs getting less respect (and emoluments) than white collar work. This results in the education system not being tuned to match with the skillset requirements of industry. Private entrepreneurs in education do not invest in vocational institutes with links to STEM subjects but do so for a major fraction of engineering colleges devoted to B. Tech courses. A large fraction of these engineers get jobs doing undervalued tasks. Higher education and human resource development in general has been neglected by most governments of all political hues. On top of it, the Indian constitution has restrictive clauses as education is a concurrent subject with both state and central governments being responsible for it. Widespread heterogeneity in technological progress within the country is also partly linked to different levels of contributions to education in different state government budgets and also differential assimilation of technologies.

Besides STEM HE, the other two legs on which hi-tech industrial manufacturing rests are R&D and industrial enterprises. Indian R&D expenditure considered sector wise indicates a big lacunae despite good progress on some fronts. India has the basic prowess in most of the medium and some high technology sectors. There are major sectors of significance, both from strategic perspective as well as jobs, like telecom, materials etc. which are neglected by both private and public sectors.

Concurrent to redefining scientific research and technology development for the sake of hi-tech industrial manufacturing is also the necessity to examine the way enterprises conduct. India has a large number of small & medium enterprises (SMEs) and a majority of them employ a small number of workers to avoid the glare of local labour laws. Most of them have poor productivity and are devoid of any R&D. This is when they contribute substantially to the Indian GDP and exports and are responsible for large fraction of jobs. Also, a large majority of big companies does not do any research to improve its productivity and prefers to cut corners on either the quality of product or wages. Neither do they follow the best management practices. In all international rankings, India is found to be among the countries with maximum potential for improvement [2].

These are a few examples concerning India which may throw light on what needs to be done to bridge the divide between developing and developed nations as far as technology development for manufacturing is concerned. These indicators are very gross and hide the real variables. Each country has to be understood in terms of its detailed dynamics of processes, cultural traits, organisation of STEM HE system, scientific research and of course nature of industrial development. India can be a good example as well as a partner country for various reasons for promoting scientific and industrial progress in many developing nations.

India seems to be showing signs of realisation about the necessity of such efforts. An example is in its proposal on the International Solar Alliance (ISA). On the sidelines of Paris Climate Conference, this new platform was an Indian initiative to facilitate best harness of solar energy for all solar rich nations lying within Tropic of Cancer and Tropic of

Capricorn. Such a move was long overdue. The fact that it came from India showed its maturity. So far 121 countries have joined this effort and a large majority of these countries are developing, while the remaining few are developed. India has thus attempted to be a bridge between developing and developed countries. India has raised its own target of solar power from 20 to 175 GW by 2022. This is in response to the crisis related to climate change and the related challenge of meeting international commitments on this. This alliance offers a contrast to the alliance of developed countries for the sake of nuclear and particle physics Centre for Experimental Research in Nuclear Physics (CERN) involving high energy accelerators and International Tokomak Experimental Reactor (ITER), aimed at ultimate fusion-based clean energy source. These are very costly projects of futuristic value and call for even developed and rich countries to join hands. These are costly monoliths, but they might have yielded lots of spinoffs to many of these countries helping them to maintain global technology lead. The true test of ISA will be how it influences the course of technology development that impacts developing countries. These technologies have the potential to be of game changing nature with focus on decentralised deployment, inclusivity and empowering poor and remote communities.

Besides climate, many other disruptive technologies also call for major pooling of efforts related to adapting and localising high technologies, particularly those with disruptive nature, as well as addressing local problems with their help. There are various facets of the problem and the development of appropriate human resources and skill base is only the beginning of the solution. This may also involve managing technology disparities between communities, adaptation and dissemination of latest technologies as well as opening up new innovation and technology frontiers relevant to local needs. This may be a bigger challenge than that of solar energy and its complete nature is not easy to comprehend. The more populous and bigger the country, greater are its problems. Thus, India itself being populous with a favorable potential demography can again be the axis of research and development activities in forming a consortium of developing nations for the sake of joint efforts to develop and deploy distinctively several of these technologies. This is a multi-dimensional problem in nature and India itself may have to reinvent its R&D programmes, both public and private, and reorganise STEM higher education and promote relevant industries, particularly SMEs. In most of these efforts, India could partner its development efforts with developing countries. They can together be stimulated to harness a collective talent pool of all these nations.

It is worthwhile to examine in detail the inhibitors limiting the growth of Indian R&D – hi-tech manufacturing ecosystem as despite tremendous existing limitations, its potential is also huge. In the following work, an attempt is made to do the same without the expectations of exhaustiveness. How this can benefit several other nations in similar category is also touched upon. This way, this work seeks to address researchers and developers in developing countries like India on how to be more responsive to problems rooted in their soil, be it related to technology development or deployment or developing human resources. It also aims to address research managers to envision research that promotes inclusive and sustainable development, avoiding aping of developed countries and spreading a culture of respect for genuine innovations. They need to reach out to other developing countries with similar situations for pooling of efforts for worthwhile joint R&D projects. Funding agencies and policymakers have a significant role to play by providing funds or facilitating (manpower and other infrastructure included) when and where they are required for the sake of growth of innovative R&D addressing problems of development that respects originality rather than simply follow models that support international trends, and yet support basic sciences for encouraging nucleation of innovation and culture of strategising for the future. This work also addresses business managers to participate in the process of indigenous R&D to marketed product cycle growth in the developing environment and go beyond the model for marketing licensed technology products. It also hopes diplomat-scientist or scientist-diplomat (a tribe to grow) to bring about many more progressive international alliances related to latest technologies appropriate for developing economies on the lines similar to that of the International Solar Alliance.

Chapter 3

Science, Technology, Innovation and Manufacturing: Indian Landscape

Introduction

Indian manufacturing has caught the attention of its planners for the challenges it faces and the opportunities it provides. There are a host of problems which are related to regulatory regimes, demand economics, impending reforms in laws and many others. The leadership may be serious, but manufacturers perceive it to be a slow-moving machinery. There is plenty of interest and debate about several related issues. But evidently, the bigger hurdles are not letting the country see bigger opportunities. These opportunities are those related to the game changing potential of scientific and technological intervention. At present, technology development and its appropriate deployment has a lot of unharnessed potential to change the scenario in a big way. A large part of policy making is enamoured by new digital and financial technologies and related services while ignoring hardware-based high technologies. A large amount of institutional science and technology infrastructure exists, but is more or less divorced from the manufacturing sector. In debates of this kind, the political leadership, economists and bureaucrats tend to dominate while scientists and technologists lie low. There are reasons related to their own internal ecosystem which inhibit them from intervening and making an impact. This is unlike what happens in most of the technologically advanced nations.

MAKE IN INDIA VS MADE IN INDIA

Indian manufacturing has been constantly lagging behind other sectors of the economy (like services and agriculture), with its share in GDP growth being around 15%, which is disproportionately low when compared to that of several fast-developing economies like China, Korea, Malaysia, Thailand etc.. This together with the huge demand for employment prompted Government of India to set up a National Manufacturing Competitive Council in 2011 under the direct supervision of the Prime Minister's office. In 2013, a new manufacturing policy was announced. This policy aimed at setting up investment-cum-manufacturing zones in the country. Evidently there are a lot of challenges in implementing this policy which are related to land, labour, power, regulatory regime etc. There are other issues which are significant but get much less attention. While realising low level of technology and depth in manufacturing, the former plan panel had suggested the route of FDI and joint ventures with foreign partners for acquisition of technology [7]. These routes have so far proved to be of doubtful value with lots of hurdles and pitfalls of their own.

The new government, soon after coming to power in 2014, launched a major initiative called 'Make in India.' This lent further urgency to earlier initiatives. This move had multiple goals in mind — to create more jobs, save precious foreign exchange and import higher technology. It selected 25 sectors for emphasis on job creation and skill enhancement. Defence sector manufacturing is separately considered to be one of the major areas of interest as this sector imports heavily. This programme has some laudable goals and is very aptly positioned, but there are two very pertinent questions which may be faced at a later stage. Firstly, how many jobs (quality of jobs) are likely to be created and secondly, what type of technologies will be transferred by foreign players. Economists do talk about production efficiency, which is intimately linked to developing, importing and adapting latest technologies. Can a nation of India's size sustain such missions based purely on import of technologies? What are the possibilities of developing and adapting new technologies indigenously? It has been realised time and again in the case of strategic technologies that imported products have caused more trouble over a period of time. This will start becoming equally significant for non-strategic but commercially important cases as well. Thus it will become imperative for the R&D sector to develop indigenous solutions (technologies) to deal with local problems.[1]

There is another doubt raised by many about the extent of emphasis to manufacturing. Why a developing country like India, whose economy is dominated by agriculture and which has a fast-growing service sector, should be interested to promote manufacturing? And what type of manufacturing it should be interested in? Largely populous developing countries need manufacturing for providing employment to their millions, besides other goals. There are several shades of opinion prevailing in the country. There are many planners who bank on service sector growth rather than manufacturing as the world is already suffering from the overcapacity offered by China. Internationally too, many experts have opined that most of future growth in employment will come from the service sector. There are others in the industry who see in growth of automation and robotics an opportunity to increase productivity and bypass labour laws. The moot question is: will these robots be imported or developed indigenously? A reasonable opinion is that the manufacturing sector cannot be ignored and both robotics and cheap labour may need to prevail concurrently. At some stage a law may have to be enacted regarding robotic deployment substituting for cheap labour available in abundance. The urge for achieving better efficiency and productivity in manufacturing is understandable, but India cannot ignore the social compulsions of providing jobs to its millions. It is not realised that this is a

---

[1] Sustained denial of advanced technologies by developed countries to India is quite common and, in some cases, sabotaging of indigenous efforts by all means is not ruled out.

challenge more for technologists than for economists. They have to reconcile these seemingly contradictory demands of increasing efficiency and employment at the same time.

In the context of this debate on productivity versus employment, one has to identify the roots of poor productivity of Indian industrial sector. In an India enterprise survey, it was found that a major fraction of manufacturing is done in the informal sector [8]. It has about one fourth of capital and about three fourth of workforce. In India, micro and small enterprises have low level of technology and education as well as skill levels of employees are poor. Figure 5 shows the contribution of MSMEs to GDP growth and employment in a few manufacturing nations [9].

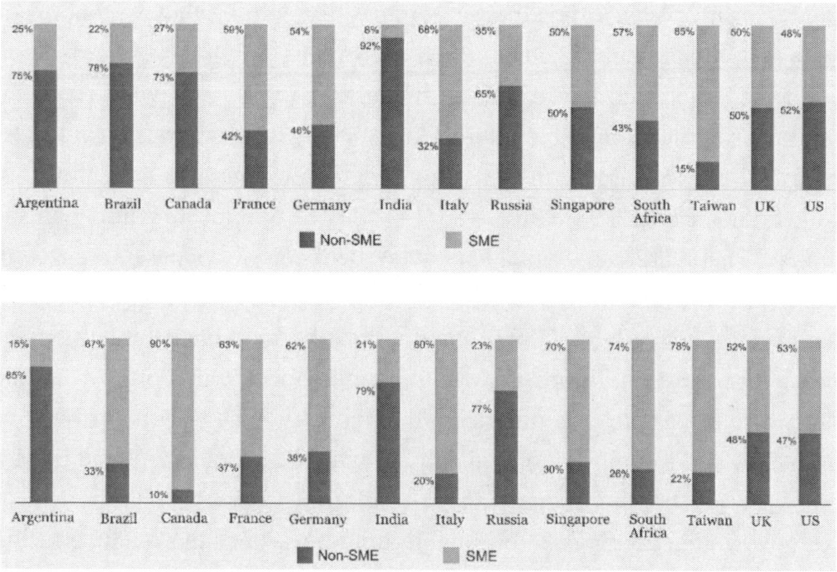

Figure 5. Role of MSMEs (and non-SMEs) in their percentage contribution to GDP and employment in a few countries. The top part of the figure shows contribution to GDP and the lower part is related to contributions to employment.

It can be seen that Indian micro small & medium enterprises (MSMEs) employ larger workforce than many countries, but contribute less to the GDP indicating poor productivity. It is well acknowledged that improving productivity and efficiency of this sector can have a more inclusive and significant impact. As a consequence of value addition, these enterprises can become globally more competitive. Again, it is not realised that technology dissemination even from within the country has the potential to turn around the scenario. Educators, innovators, scientists and technologists can in principle play a major role to alleviate this situation provided the right mechanisms and mindsets are generated.

It is well known that all developed countries manufacture high technology items efficiently and are able to provide high-wage jobs to their citizens, while most developing countries manufacture low or intermediate technology products implying low wage jobs. If India has to generate quality jobs (not just jobs!), it will require well-directed efforts towards design, development and manufacturing within India. It will have to be more 'Made in India' than 'Make in India.' 'Made in India' is expected to aim not at manufacturing alone, but also at developing more deep-rooted capabilities right from the stage of conceiving novel products (and processes) to their designing, development and manufacturing. This goes beyond the earlier paradigm of self-sufficiency in certain chosen sectors as it seeks to focus on whole of manufacturing for the sake of better employment and for serving the necessities of domestic consumers or global supply chains by covering most layers of technologies. Scientific research and technology development in the country has to reinvent and reform itself to achieve this goal.

RESEARCH AND TECHNOLOGY: MAKING RIGHT CHOICES

There are often questions about what kind of technologies a developing country like India should develop. At what stage of emergence of a new technology on the global scene, R&D efforts should be launched within the country and by whom. Similarly, at what stage full-fledged and competitive manufacturing can be initiated and by whom. An equally important issue is how these technologies should be deployed and exploited for maximum impact. Two factors which usually dictate the technology level of manufactured products are domestic market demand

and exports. It is also influenced by the domestic level of technology development and skill level of the workforce. In the absence of indigenous technologies for products in demand, there is no option, but to resort to imports. Technologies required for internal demands and that for exports may not always converge. For example, it is reported that in most OECD countries high and sometimes low technology is employed for exports, but domestic markets demand medium range of technologies [1]. In India, the domestic market traditionally has been dominated by low price and consequently low quality and technology of products. But of late, there has been a marked change and demand for quality products and interest in high technology products have been gaining ground. One may say the domestic consumer requirements cover a wide spectrum in terms of level of technologies, with India being a large and heterogenous country[2]. India has to calibrate its choices for indigenous technology development and manufacturing accordingly. Making products for exports requires a higher level of capability and success in this for a particular sector indicates maturity of that technology in the country. Entering international supply chains is not so easy and may require Indian manufacturers to start with entering at the lower end and with components and sub-systems then gradually move upwards. India does have a segment of research-technology developers which can produce high quality products. For example, Indian strategic sector has islands of fairly skilled workforce and the sector banks on high quality and high technology. If India can ensure pervading this to other sectors of manufacturing, it could be a big gain. For achieving this, there is need to evolve appropriate alliances and mechanisms. Traditionally Indian industrialists seek licensed technology from abroad assuming that to be the most economic and competitive route to profits. It may require a lot of efforts to persuade them to interact with Indian researchers and technology developers and evolve suitable mechanisms for manufacturing based on indigenous research.

---

[2] The increased interest of Indian consumers in high technology products is evident from extensive sale of high-end products from companies like Apple Inc (plus many others of this category) and interest of some of these companies in manufacturing products in India.

There is wide range of technologies, skill levels, demands as well as products in the Indian domestic scenario. There could be products which are of high technology and high value but are needed in low numbers, mostly in research or strategic applications. There could also be products with medium or high technology like medical equipment, generic drugs, light emitting diodes (LEDs), smartphones, laptops, solar panels and therapeutic lasers and other medical electronics items which could be required in large numbers. The defence sector needs a wide range of technologies in large volumes. There could be frontier technologies related to artificial intelligence (AI), autonomous vehicle, electric cars, energy storage technologies and so on. Each of these categories demands R&D efforts of different kind. But products based on them may soon be in demand in large volumes.

Many of the developed countries have over the years achieved specialisations in certain categories of technologies and designs. For example, the Germans are known for cars and other transportation equipment, Americans and Koreans for ICT-related products, Swiss for precision machine tools, French and Italians for textiles and apparels etc. Different countries tend to place different levels of emphasis on developing and deploying nascent and more disruptive technologies like internet of things, robotics, artificial intelligence and 3 D printing. All developed countries with sufficient capital tend to invest in them at an early stage. For example, robotics may be of interest to developed economies for assembly line manufacturing as they may like to beat low wage manufacturing in developing countries. Artificial intelligence could be of interest to Indian IT industry for substituting low skill IT jobs for the sake of retaining competitiveness. Many of these technologies in developing countries like India do not evoke any interest at an early stage of their growth and this often results in the country as a whole missing the technology growth bus. Autonomous vehicles, robotics and AI are intimately linked to software and as a nation, India has the opportunity to leverage these capabilities by encouraging R&D. The policymakers have to go beyond the usual 'pragmatic' approach (offering limited resources excuse!) and encourage R&D in these areas so that the country can harvest it later. It is beyond imagination of many policymakers to appreciate possible applications of many of these technologies in the Indian context. An example of additive manufacturing and 3 D printing could be illustrative as it is of interest for more economic production of very specialised products in a much shorter time avoiding wastage of precious materials. These attributes are of great interest to any developing country. Possibilities like these plus dispersed deployment empowering small scale producers or consumers should be attractive to a country like India. If this pervades, Indian MSMEs and young startups may upgrade their productivity and diversify their potential. While R&D and technology development needs encouragement at an early stage of growth, manufacturing can be decided upon at later stages. These high technology sectors can create a number of high skill jobs. Before deployment of new technologies, appropriate preparations of scientific and technology manpower which can help assimilate it smoothly is crucial for wider dissemination and impact. There has to be appropriate synchronisation of efforts between different segments related to human resource development, R&D and manufacturing for the sake of maximum impact[3]. Such coordinated actions are usually wanting in most developing countries like India.

Another example of an R&D area where strong emphasis is expected from countries like India is that related to renewable energy. Presently this R&D sector is not getting sufficient attention and there is disproportionately large emphasis on deployment. For example, with India's ambitions for large scale deployment of solar power, it could stimulate R&D in several related directions like development of efficient solar panels and energy storage. This has to be done in public sector R&D as there are major international competitors with deep pockets with whom the Indian private sector may not like to compete. There is an opportunity to promote related R&D through the auspices of ISA. A major boost to

[3] Robotics in India is not growing at a pace it should have progressed, in light of the possibilities to leverage its IT strengths. Besides its integration with application areas like autonomous vehicle, surgical operations, internet of things (IOT) for operations in hazardous situations, deep bed mining, radioactive areas, toxic chemical/biochemical plants, defence-related operations, cleaning septic tanks and many others have not happened. The delay in large scale development/deployment could prove costly missing of buses. This is linked to lack of trained workforce with requisite knowledge base and skills.

certain technologies like these can thus be influenced by a country like India by leveraging its economies of scale. A large number of developing countries collectively need to exercise their potential to influence the course of technology development by way of market influencing and distinctive deployment. This is possible only if they put their internal ecosystem in order.

EVOLVING SYMBIOSIS OF RESEARCH AND MANUFACTURING: SPECTRUM OF SECTORS

'Made in India' could be a challenge for researchers, developers and innovators as well as for those who manage research in the country, but it has ramifications beyond the R&D sector. It is a complex and multi-dimensional goal. Science, technology and innovation system has to respond to techno-socio-economic situation in the country. Both the technology scenario and socio-economic situation of the country are changing constantly and thus continual review of plans becomes necessary. There is need to carefully assess what scientific research and development can be done in this context. India spends only about 0.7% of its GDP on R&D and about 70% of this is spent in the public sector. Out of this, 60% is spent on sectors related to defence, atomic energy and space. About 8% is spent on industrial research, 4% on agriculture, 1% on medical research and about 4% on all the R&D in universities. More importantly, all the major spenders are confined to work on their specific mandates and silos. The statistics on funding details are for broad guidelines and also change from year to year. India has 5.28 hundred thousand researchers out of which only 53% were involved in active research and rest were in auxiliary or administrative services (in April 2015). This included in-house researchers of public and private sector [5]. Indian expenditure on R&D was around 2.7% of global share and its publications more than 3% [5], but its working patents record was very poor. There is a lot of distortion of statistics about distribution of R&D expenditure and patent record between the government and private sector as in India, MNC R&D centers are emerging as a new category of a major R&D spender group. MNC R&D centers are spinning a large number of patents (instead of publications) as compared to Indian entities. The scientific research output may not be considered bad, but the national capability to utilise its R&D for manufacturing sector and job creation may be questionable. Good quality well directed research, in principle, should have good employment generation potential. This may seem not acceptable to many in the R&D community. But if managed and harnessed properly, every good quality researcher should be able to cause generation of ten new jobs or train at least ten more researchers. Strategic R&D sectors generally cannot impact production of goods for mass consumption and exports. Thus, presently their direct contribution to manufacturing is not of much consequence and they happen to be major users of R&D funding. But they can contribute substantially if most hi-tech manufacturing required by them can be done indigenously. They can also contribute to dissemination of knowledge base and human resource development. How can countries like India promote the R&D sector to go beyond rigid mandates, stimulate hi-tech manufacturing and start contributing more actively to say, missions like 'Made in India'?

There is need to reassess all relevant aspects related to R&D in light of these additional tasks and how India should make use of its limited scientific manpower resource. Thus, choices in technology development, manufacturing and R&D directions have to be viewed collectively. The strategy paper on manufacturing by the Planning Commission (2012) discusses problems faced by each sector [7]. For the sake of present discussions, most of the sectors can be categorised based on their linkage to the type of R&D contributions possible. These sectors could be, for example, traditional manufacturing industries like textiles, gems and leather or modern manufacturing industries like pharmaceuticals, automobiles, chemicals. All industries related to defence, shipping, railways, aerospace, nuclear, telecommunication, computers etc. are clubbed as high technology, this being the third category. Here, shipping and railways are also categorised as hi-tech against standard international convention as India is still heavily dependent in these sectors on imports and performs little R&D. Lastly, the fourth category is that of digital manufacturing industries which are critically linked to computers and internet such as IOT, AI, 3 D printing and additive manufacturing. Some of these are supposed to usher the world disruptively into the fourth industrial revolution. These technologies are of special interest to India as its IT prowess can be good nucleating factor. Also, it may involve extensive job destruction as well as creation. Each of these technologies and related industries has different set of problems in which R&D efforts of specific kind are needed.

All these industries can also be differentiated in terms of their employment potential, impact on environment, efficiency of resource utilisation, productivity of man and machine employed etc. Many such goals have to be suitably targeted while R&D links to manufacturing are established [10]. For example, textile industry has been a major bulwark of Indian industry and is valued at about 105 billion dollars with about 14% of total industrial production, 15% of total exports and a contribution of 4% to the GDP. More importantly, it employs about 45 million workers. Combined with indirect employment, this figure goes up to 100 million. It has progressed in terms of volume mainly because of the use of genetically modified (GM) cotton, as India has become the biggest producer of cotton by surpassing China. But despite this, India's share in the global textile market is a small fraction of that of China. This is linked to poor value addition in India. In addition, this industry does contribute to pollution by way of processing, dyes etc. Indigenous R&D may help in modern machinery, better automation, eco-friendly dyes, productivity and modern as well as economic quality control, internationally acceptable designs and more importantly percolating it to smaller units. For many of these goals, the Indian textile industry depends on foreign sources. By improving productivity, it may not necessarily improve the scope of employment directly. Real gains could happen if most related R&D and technology upgradation and development of modern machinery required are carried out from internal resources. Compare this to the IT sector, which has over 125 billion dollars of exports, 41 million dollar of internal market and 3 million employees. But the quality of an average job is much higher [10]. This does not contribute to pollution but needs to go up the value chain and develop IT-related hardware. The e-waste problem has not yet caught sufficient attention but may require R&D for efficient recycling, developing IT products and contributing in budding new age technologies like AI, machine learning, IOT and robotics industry. It is attempting to bring in automation which may impact low skill employment. This industry also imports a huge amount of IT-related hardware. In contrast to this, the pharma industry, which has also grown quite rapidly in recent times, has exports worth 27 billion dollars, with about 0.5 million employees of high skill variety. This industry has smartly exploited reverse engineering but has so far failed to make a dent into the area of new chemical entities as well as clinical trials, where challenging R&D is needed. A nascent sector which has made significant progress is automobiles. In terms of sales volume, it has achieved the fourth rank in the world and is growing at 6% on YOY (year over year) basis. Directly as well as indirectly, it is supporting millions of workers. This sector is active in R&D more than many other traditional sectors. With rapidly changing scenario of electric and hybrid vehicles, a serious upgradation of R&D efforts may be necessary. A particular area of neglect in India is driverless vehicles because of the questionable presumption of it being a poor match to Indian market demands. Another promising area could be research related to sector of automobile safety at competitive prices.

Some of the traditional sectors too demand R&D. The gems and jewellery sector contributes about 7% to the GDP and 15% to exports and employs more than 4.5 million people. Its competitiveness could be enhanced, if the players can be provided low cost indigenous processing and testing equipment and machinery. It is important to remember that besides present-day employment in a sector, one must consider the potential rate of growth of employment vis-a-vis productivity also. This is dependent on R&D performance and its linkage to manufacturing. Similarly, food processing and the R&D related to it is an important sector which has not got the attention it deserves. About 70% of India's workforce is engaged in agriculture or food processing, directly or indirectly. The overall food market is worth about seventy billion dollars and about one third of the value comes from food processing, which in large measure is performed by relatively small units. There is an estimate that about one fourth of foodgrain is wasted because of lack of appropriate technology for storage and handling and processing post-harvest. It is obvious that the scope for technology intervention is huge and the government is banking heavily on foreign sources as FDI limits have been raised to 100%. Dependence on imported technology may not fill the gaps completely and will be slow to come by. Indigenous R&D can surely play a role if various technology developers work together. This may require going beyond the government institutions presently expected to work on this. The development of this sector could also involve organic crops, GM foods and irradiation of foods for preservation and many such other technologies. Indigenous technologies and consequently manufacturing will serve twin purposes of improving efficiency of food utilisation and creating more jobs. These technologies will need to face competition from MNCs. The Government of India has initiated a new programme of AGRI UDAAN to encourage innovation, mentor startups and bring them in contact with investors. This has good potential if well dovetailed to appropriate R&D.

Lack of R&D in any sector leads to poor value addition and underselling of precious resources. As it happens in the case of textile industry which exports a large amount of cotton instead of ready-made garments, a similar phenomenon is noticeable in the case of precious mineral ores. This is indicated by the fact that the average share of material costs in production is about 80%, as compared to 40% in the case of a technologically advanced nation like Germany [11]. India's overall resource productivity is considered about one tenth of that of Germany. There are other reasons for protecting and making best use of national mineral resources. Many of these precious metals may be in short supply in future or could be even restricted in trading because of strategic considerations. America has enacted a law recently to ensure prudent use of its mineral resources. A careful scrutiny of various indigenous minerals based on titanium, tungsten, rare earth and atomic minerals (most of them of strategic value) demands complete metallurgical processing (including byproducts) to prevent unscrupulous exports and underselling of India's mineral resources.

There has been a major movement of population from rural to urban areas in most developing countries. This has caused chaos-like condition in many metros all over the world. Most significant perturbations have been those related to environment, transport and healthcare besides many others. A United Nations report predicts that in less than 25 years, the urban population of India will exceed that in rural areas [12]. One can notice the rate at which the magnitude of attendant problems may increase. India has some of the most polluted cities in the world. Besides intra-city transport, there are major intercity transport problems also. These have major implications for citizens as well as manufacturers related to movement of goods — both by ways of logistical cost and restrictions. The Government of India has announced policies on smart cities and industrial corridors. Both of these face major finance, regulatory and infrastructure-related hurdles. As usual these are problems which overwhelm those expected to implement these tasks. Some of these projects run into billions of dollars. But they spend a minuscule of the same on innovation, scientific research and technology development related to these problems — most of the times foreign companies are coopted for the purpose. But more important than finance alone is other investments like human resource and improving linkages between research, manufacturing and real-life solutions of problems. There are many other sectors like railways, aviation, telecom and infotainment where efforts related to R&D are pathetic and expenditure is mostly controlled by bureaucrats or the bureaucratic scientific leadership. This mindset of inadequate attention to innovation and R&D is causing a huge loss. The same attitudes run through majority of developing countries. Such inaction and indifference prompts developed countries and their companies to take advantage of their big lead.

As can be seen from the examples cited above, R&D could significantly add value to many industries in various situations. This goal may not always demand invoking cutting edge or frontier technologies or new patents. Many of India's manufacturing industries are at a stage where they may not need frontline knowledge. This does not imply banking on off the shelf importing of equipment, for the sake of technology upgradation. Much of the concerned knowledge is not new to the world and not even to the country as a whole. As mentioned earlier, a World Bank study had pointed out at the existence of large variation in productivity and related knowledge base within R&D and manufacturing entities (services sector included) in the country [8]. This implies that mean productivity of manufacturers in any sector is much smaller than peak productivity achieved (or possible) within the country — this itself of course is much smaller as compared to what is best achieved globally. Even sharing of this knowledge base which could include technology as well as management practices between various segments of Indian manufacturing and R&D entities can boost productivity several fold. This requires imaginative handling preferably by technologists and domain specialists (and not bureaucrats alone). The usual modus operandi of providing support is to provide funding and allow imports of upgraded machinery. This rests on an assumption that R&D can be made available off the shelf. This is hurting the country at large. If entities within the country can help upgrade technology of these MSMEs, which in the case of countries like India are in the domain of possibility, this can have major impact on their overall competitiveness. The most pragmatic step required is to have technology scouting, hand holding between researchers/ technology developers and manufacturers and bringing down such exercises to more specific levels of R&D-based problem solving. The Government of India has recently initiated certain measures like technology acquisition fund for value addition and upgradation of technologies in MSMEs. Benefits of such schemes will multiply manifold if technology upgradation is performed with the help of internal sources (who exactly?). It is limited imagination which impels planning for technology upgradation resulting in import of machinery with higher capacity or capability.

India must evolve its own model of indigenous research and deployment of technologies in many areas which is commensurate with its requirements. Many a time, combining the dynamism of private sector for faster deployment plus broad base of S&T development of public sector labs can be quite fruitful. As mentioned earlier, the Indian pharma sector is a very clear-cut example of the same. India Brand Equity Foundation attributes growth of Indian pharma sector to appropriate public-private partnership, increased penetration of healthcare facilities in non-metro cities, involvement of multinationals in setting up facilities in the country, international standard of pharma education and rapidly developing R&D. [13] Indian farmers adapting new varieties of crops following suitable preparatory exercises is another example. Thus, developing the right kind of skills and higher education contributes in a major way to diffusion and widespread deployment of technologies.

A noticeable related point is the significance of role of public-private partnership in R&D, particularly when big investments and higher level of knowledge base are involved. The bottom line is that Indian scientists and technologists have got accustomed to work within the four walls of government labs and this has restricted diffusion of newly developed technologies. Thus, many a time technologies developed here have not reached their probable beneficiaries. The existing mechanisms of technology transfer seem like soulless efforts with less than optimum yield. Effective and widespread dissemination of technologies is not easily done in the framework of government labs and may call for some more enterprise on the part of Indian researchers with adequate support from the government. The new Indian policy of involving private sector in manufacturing of defence-related items could, if implemented properly, become an example in future for many other sectors where government researchers and private manufacturers can work together.

Special Case of Strategic Sectors

On the other end of spectrum is the concern about growth of strategic sector, high technology and related big manufacturing. Defence imports for India are expected to be about 250 billion dollars in the next five years and a similar figure is expected to be for electronics sector (there are several estimates). These hi-tech sectors can alleviate the bigger stress on the economy in the form of its oil import bill. Thus defence, nuclear, aerospace, communication, computers and electronics need to be the focus of attention and development of related technologies will be both a challenge as well as an opportunity. Private industries are not normally involved in these sectors (except consumer electronics). Technologies are not easily transferred by foreign manufacturers in these sectors and the demand may keep growing. There is no universal solution to this and it is quite doubtful that FDI, joint ventures etc. can solve the problem in all cases. Even if it happens, in each case it must be weighed against the cost and efficiency involved and compared with long term solutions like indigenous development of technology.

Defence and security technologies are of such nature that many armaments and related products can be imported, but technologies are not easily transferred or are priced heavily. In this context, one need to understand the perspectives of traditional supplier/partner countries in these sectors (Russia, Israel, European nations and lately America). Most of them had been mainly suppliers, except for a few co-production projects. Normally co-production has limited benefits and it leads to assembling and repair work by the local partner and further leads to perpetual dependence by way of spare parts. The manufacturing public sector units (PSU's) also continue to remain predominantly assemblers or system integrators. Some of the exporting countries may agree to co-development or co-production of limited range of sensitive technology products for their own reasons of capturing big markets like India. Even in co-production projects, a technology junior partner cannot hope to be in advantage. Recently, India integrated its Brahmos cruise missile with Sukhoi aircraft. Indian scientists and technologists with their expertise could do it at a fractional cost of what was quoted by its partner country, Russia. Such examples illustrate cost benefits of indigenous R&D. Similarly, America, despite having an agreement with India in the form of defence trade and technology initiative (DTTI), does not easily pass on critical technology of relevance. Most of its frontier defence science and technology get nucleated in government labs and it does not share even with its closest allies [14]. There is another category of mature American defence technology platforms and products which are produced in their private industrial conglomerates. These industrial players have little interest in co-development with India for reasons pertaining to a wide technology gap between them and Indian entities. For co-production, they also expect 100% FDI and assured orders. Thus, the cost of technology import through FDI and co-production is not small and can be acceptable only in few specific situations, and indigenous technology development needs to either compete (or substitute) or supplement. In the face of such a situation, another route suggested for the purpose of getting technology transferred is of exploiting offsets, capabilities to assimilate. This is also dependent on its indigenous R&D strengths, status of STEM HE and skill levels available within the country. Also, any transfer of technology under this route will require agreement of government of the exporting country. There is a recent example where the Republic of Korea has been denied a few technologies developed by the supplier, Lockheed Martin (in lieu of offsets), at the instance of the American government, even when Korea is considered to be a close ally of America [15]. Besides all these hurdles, offsets are prone to corruption in countries that aren't particularly sanitised. From the observations mentioned above, it is evident that there is no easy route to going up the technology proficiency ladder in all high technology sectors. Cryogenic engines for space sector, super computers and steel plants in an earlier era are another set of similar reminders to Indian technology developers. Continued denial of high technologies is a regular feature and reasons are not far to seek. The bottom line is that efforts for indigenous R&D for manufacturing in high tech, strategic, and now more importantly, commercially significant sectors are imperative.

The Indian defence sector is a case where R&D institutions like DRDO and defence PSUs have together been sharing the major burden of both R&D and indigenous manufacturing. This has been useful with some limitations. This has stunted progress in "cutting edge" R&D as well as speed and scale of deployment of new technologies. However, recently, the wisdom of involving the private sector, at least for manufacturing, has dawned upon the policymakers. By sharing some of their development burden with private sector, Indian defence R&D can be liberated to move in more worthwhile directions of developing frontier technologies. By sharing long term perspective planning (LTPP) with private sector (including MSMEs) and committing to sure orders for a well-developed technology, this relationship will deepen and ensure proper growth of defence-related manufacturing as well as R&D. While sharing long term plans in the defence sector, the government can share the risk of development expenditure with private sector in initial phases. As the private sector matures and starts making profits, this burden can be gradually shifted to them. The most distinctive feature of defence-related manufacturing is that there is a huge demand waiting to be tapped. Foreign original equipment manufacturers (OEMs) and Indian private sector areas of manufacturing have to be suitably demarcated without avoiding competition. The Indian private sector should be encouraged where the government is not able to lure foreign OEMs to bring technology. The Indian private sector may find it attractive to collaborate with Indian researchers for boosting their prowess. The recent decision of the Indian Cabinet to accept model of strategic partner in private sector for various segments of defence production can be only a welcome beginning. This model needs some tweaking as it again emphasises more on finance and procurement (concept of strategic partner etc.) related issues and less on ideas, innovation and research components which are not always hostage to capital. These shortcomings have to be overcome suitably.

The other segments of Indian strategic arms are nuclear and space sectors. Each one of them has problems. Nuclear power, because of its green nature, continues to be of great interest. In India it is expected to grow on two parallel and yet complimentary paths. In the first one which is totally indigenous, this involves nuclear power plants in the three-phase programme of DAE involving pressurized heavy water reactors (PHWR), breeder reactors and finally thorium utilisation. It involves extensive R&D, particularly related to thorium. India has to do this R&D in isolation as there are a few foreign partners willing to get involved. Thorium technology is crucial to India as it owns large reserves of thorium and suffers from uranium shortage. Few collaborations of DAE with USA and

Germany for getting better access to high current accelerator technology (aiming for novel ways of thorium utilisation) hopes to serve the purpose in a limited way. But manufacturing for its existing designs of nuclear reactors (U-Pu systems) can progress irrespective of such R&D. The Union Cabinet's recent decision to sanction ten 750 MW indigenously designed and proven technology of pressurised heavy water reactors (PHWRs) is a good beginning which banks on indigenous R&D.

The second route is foreign nuclear power plants from Russia, France and USA. India started getting uranium from various international suppliers after the Indo-US nuclear deal. Widespread deployment of nuclear technology for power requires foreign suppliers. But the problems related to nuclear liabilities have to be sorted out and now may be coming to an end with the pool of 12 insurers entering the scene. Also, Japan has agreed to supply various critical components for American and French projects. Most of these reactors will require a lot of non-nuclear manufacturing which will accrue to Indian suppliers and range from 40-50% of total expenditure/investment. There was a news report that suggested that American supplier Westinghouse may offer its technology and will prefer that the rest of the task is completed within India. This demands high quality work which several Indian manufacturers are now trained to do with their background of working for Nuclear Power Corporation of India Limited (NPCIL). Besides the liability issues to be settled with suppliers and insurers, India must factor in the additional costs accruing due to waste disposal and related environmental issues, allaying public fears before India can move fast in the direction of boosting deployment of nuclear technology for power generation. This again involves frontline R&D on nuclear waste disposal. Indian nuclear R&D organisations should offer accreditation and quality certification services to local manufacturers so that they can plan manufacturing and permissible exports in the same sector. Such possibilities will brighten once India becomes a member of Nuclear Suppliers Group (NSG). This should give additional boost to private Indian manufacturing in the nuclear sector.

The space sector is growing rapidly, having acquired capabilities in very diverse sectors based on a variety of launch vehicles. India has perfected the technology of low earth orbit satellites even in the higher weight category for a variety of applications. Its record is very good by international standards. It is yet to perfect Geosynchronous Satellite Launch Vehicle (GSLV) — the advanced versions. Its commercial arm, Antrix Corporation, is offering services in satellite launching, remote sensing and weather forecasting, and will soon offer services on navigation and cartography applications. Currently there is an increasing demand for small weight satellites. India is yet to develop such a launch vehicle. It can be seen that India has fairly good capabilities which have got some good publicity with the excellent and frugal Mangalyaan mission and launch of more than 100 satellites simultaneously. But its capacity building still requires work. While Europe, USA and Russia command about 80% of the world launch market, India's share is around 2-3%. The recent launch of GSLV Mark III based on indigenous design of cryogenic engine has great potential. But to exploit it fully, India needs to build a few more launch sites, pass on the capacity building jobs to private sector so that its hands are free for more development work on frontier technologies. Some recent initiatives of the government to involve private sector industries are in the right direction, but for this purpose, private players also have to cross several hurdles related to quality control and international legal clearances. More and more of them have to join in building quality components for most of the aerospace sector. It will be good to take cue from the Indian auto sector in going up the technology competence ladder in a step by step way. Many of the private players, including MSMEs, have already been supporting the Indian space sector. They will do better by constant quality upgradation and international level certification of their workforce. The Department of Space should take steps to facilitate this process. The Indian Space Research Organisation (ISRO) has taken some steps to involve private sector in its Polar Satellite Launch Vehicle (PSLV) launch efforts with an aim to maximise capacity and improve efficiency. This is a small step in the right direction which maybe could have come earlier. The Department of Space feels that it may take some time before the Indian private sector matures in this respect, but it is surely a seed worth sowing.

Thematic R&D for Made in India: Few Examples

Mission 'Made in India' may not restrict itself to domestic consumers and will naturally expand with time to export markets. Similarly, the demands of domestic consumers will diversify and evolve with time amidst growing competition from imports. For the sake of both these types of consumers, the manufacturing sector and related R&D will require to carve out some niche areas. Indian manufacturers may have to compete with those from other low wage countries. It may not be easy to compete on price front alone. So, some emphasis on distinctive features of quality, productivity, economy, novelty together with product design could be initial guiding principles. These are challenges for enabling R&D and related knowledge base to be harnessed for quality manufacturing. There is need to identify the main players who can bridge R&D and manufacturing, several catalytic agents involved in the process as well as overcoming hurdles. The Indian ecosystem has several inbuilt barriers to enterprise building and scientists and technologists may not be keen to venture out for technology enterprise in such a situation. There are a number of artificial and non-technical barriers to enterprise building in some major countries and India is one of the worst performers despite major efforts.

Presently, the Indian manufacturing sector continues to face a few major hurdles and some self-imposed inhibitors that stand in the way of speedy growth. These hurdles influence a large number of sectors to a different extent. Indian manufacturers as well as researchers need to be working together on a few 'super thrust' themes that have been neglected so far to overcome these inhibiting factors. Some of these thematic R&D areas are given as example below. A fair amount of knowledge base exists in these areas within the country, but will require a much more coherent and specifically targeted effort, coordination and reorientation to harness these further to impact the 'Made in India' mission. The aim here is to focus attention on some of those underlying themes which can have a direct impact on making Indian industries more sustainable, competitive and productive. These are good illustrations of how science and technology can impact manufacturing and face futuristic challenges of disruptive technologies. Most of these research themes are likely to be of significance in the near future, generating new opportunities that can have a large multiplier effect.

Environmental Monitoring and Control

Environmental controls in recent times have become one of the biggest inhibitors in the process of industrial development. How is it that developed countries are managing the two seemingly contrasting demands very efficiently? It could mostly be related to lack of objective deployment of science and technology of environment in developing countries.

Environment and climate in their own right are big issues nowadays and are being looked into by a large number of researchers worldwide. India has already proposed its plan to cut emission rates at the Paris climate meet (India has ratified the treaty too). First of all, these emission goals are disproportionately high as compared to India's cumulative contributions to pollution. Secondly, with India's expected rapid development, will it be easily possible for the country to achieve these goals? This will be true particularly in case deployment of renewable energy power is not as per the expectations. There is an additional risk of rapid expansion of renewable energy by being loaded with less efficient renewable technologies as these technologies are themselves making rapid strides. In view of many of these demands, India cannot be harsh on its manufacturing industries by way of indiscriminate environment-related curbs. It is important that these curbs are imposed judiciously and in a targeted manner. This way emission targets are likely to be achieved without sacrificing development goals. How is this possible? Through accelerating indigenous R&D related to environment monitoring and control that can play a crucial role. The recent withdrawal of USA from the Paris protocol should accelerate the resolve of developing countries for indigenous research, but it calls for extensive indigenous technology development.

The environment is not only about greenhouse gases alone. It goes much beyond. It affects public health, agricultural growth as well as manufacturing industries. There are various ways in which research can help manufacturers in consistence with environment goals. The first concern related to R&D in this area is environment impact assessment and environment monitoring related to various industrial enterprises. In India, environment-related clearance of projects has become a subject of a big controversy. On one hand those setting up industries, power plants, chemical plants, refineries or dams etc. complain of inordinate delays in getting clearances and related corruption, on the other hand there are environmental activists and project-affected people who complain that sufficient thought has not been given to issues related to long term ecological impact. It was in this context that the Supreme Court of India had ordered the government to set up a regulatory authority for pre-project review and post-project monitoring. The National Green Tribunal (NGT) so formed has to interfere in many situations, be it pollution in metros or setting up of a power plant. There are administrative and political issues related to the scope of the task at hand — apportioning of powers of appraisal, approval etc. But as experience shows, this tribunal also does not have access to detailed monitoring mechanisms and a database, and consequently the reaction is less than optimum. In the following work, some of the related technological aspects are discussed in brief.

In the absence of scientifically collected data and its appropriate interpretation, there is either delay in decision making or bad decisions are taken. This offers an opportunity to vested interests to exploit the situation by exercising discretionary powers and indulging in corruption. It is equally true that post-project monitoring, which in large measures entrusted to state pollution control boards, is a prolonged and usually unsatisfactory affair. They are not exposed to the latest techniques and technologies for the gigantic task they face. Whether it is handling of coal at Bombay Port Trust, movement of trucks through the city of Delhi, banning of diesel vehicles or shutting of mining activity in Goa, the Supreme Court is made to intervene because disagreements between opposite points of view are very common. Even many of the decisions of

NGT or Supreme Court have been questioned. Politicking and lobbying have not yet rooted out. In the absence of objective and scientific data about specific environment, wrong decisions of one kind or the other are likely. All these have deleterious influence on setting up and running manufacturing industries. The aim of NGT should not be to appear firm against any environment damage, but objectively weigh the data facts to find the most appropriate balance between environment and development. This is possible only if it is provided with detailed data and rational scientific advice.

Environmental issues inhibitors for healthy industrial growth. But if handled properly, they can open new avenues of R&D and offer good opportunities for fresh enterprises. An aspect of serious concern is the way much of unwanted pollution and lack of public awareness about its impact on the national health and consequently its burden on the national healthcare services and the economy. The World Health Organization (WHO), in its recent study [16] has concluded that one in eight deaths worldwide is related to air pollution. Developing and poor countries are among the worst affected. Some of these diseases are ischemic heart disease, stroke, chronic obstructive pulmonary disease, lung cancer and lower respiratory infections. Among these maladies are those related to indoor pollution caused by wrong fuel usage and those related to outdoor pollution to which the contributors are many but anonymous. Even though the broad causes are mostly well known, more quantitative weightages are mostly conjectures. Both types of pollution sources are causes of a huge number of patients flocking to hospitals aggravating the already overburdened public health services. As a consequence of their ill health, they contribute less to productive efforts of the nation and instead are a source of loss. A WHO study estimates the loss to national exchequer by several trillion INR, which is a few percent of the GDP of India. Pollution does play a role in agricultural productivity also. If pollution can be monitored and controlled, it shall give a fillip to food production and the economy. Efficient monitoring leads to efficient control.

There are varied sources of pollution in different geographical locations. It is not adequate to follow the generalised prescriptions for solving environmental problems in a totally different geography and other local conditions. This calls for developing countries like India to generate their own detailed database. The policy related to public issues is often dictated by the public discourse and unless the public is able to witness some immediate disadvantage of environmental degradation, it may not be easy for the government to divert major investment away from social sectors to those related to environment while not realising that this is one of the biggest social impact issues. Occasional environment-related disasters like what happened in Kedarnath or urban flooding in cities of Mumbai and Chennai, the National Capital Region (NCR) or Bengaluru are not enough to awaken public consciousness from slumber. Repeated problems related to pollution in winter in Delhi had some modicum of effect on public psyche. A clear example of heightened environment-related consciousness is politics in Germany, where the green movement has persuaded Germans to go in a major way for environment cleanup involving closing certain industries, relocating others, encouraging environment-friendly fuel and products and renewable sources of energy. The remedy may not be the same everywhere but the tools to convince the nation are the same. Thus, the first step to a major initiative in the direction of environment cleanup in India is to have suitable machinery for generating indigenous environment-related detailed database with all the technology at its disposal. While marine pollution is more obvious to all,[4] the air pollution and its effects have to be brought home with a lot of efforts on data collection and establishing correlation between pollution data and health maladies or other problems in a comprehensive manner. These databases for both types of pollution will be the first step towards ensuring compliance by polluters.

Many countries in the world have enacted clean air laws for effective environment management. Each country has an environment specific to its own conditions. India has a diverse environmental situation. India has deserts, mountains, snow, monsoon winds, vast coastal tracts, rivers and

[4] Marine pollution becomes more apparent from sale of fish (affected by pollutants) and other marine species from seas – such studies have been carried out around Mumbai and many coastal areas. Much greater effort is required for establishing the bad effects of low level air pollution, as its health effects accumulate over a longer period.

waterfronts and so on. There are densely populated cities and vast rural areas. It has fast developing industrial environment. As mentioned above, there are issues related to environment impact assessment (EIA) in pre-project and post-project situation. Because of pollution and environment-related accidents, it is imperative that monitoring of environment should be agile for the sake of faster and appropriate response in case of an industrial accident. This is challenging for monitoring and regulatory authorities. The strategy paper on manufacturing by erstwhile Planning Commission in 2012 [7] wisely recommended a switchover from a concentration-based regime to a load-based one. But there is no action on the ground again in view of scarcity of relevant data, which can pin down responsibility. Presently there is a system of simply monitoring and reporting the concentration of a particular species in a particular location and how much it is below or above the safe limit. This is followed by public outcry and blanket bans which is a worthless solution! As pollution is becoming difficult to control, a need is being felt to identify the culprits, know the extent of harm caused by each and impose suitable penalty or other remedial action. This demands a need to evaluate quantitatively the environmental load a particular activity is causing. There is a need to have "only polluter pays" principle in place. How to translate this into a technological possibility? For proper implementation of such schemes, it is important that more precise and quantitative data is generated and correlated to its source.[5] For a country of India's size, there are 703 ground-based pollution monitoring stations catering to 300 cities, as per data available with the Central Pollution Control Board of India. However, all of them do not operate at all times. The numbers are pathetically low and the number of functioning ones is even worse. More significantly, these demands may grow rapidly. Latest and efficient technologies can be a boon for the situation. Internationally, researchers are racing to build

[5] In a highly pollution prone city like Delhi, there are less than fifty pollution monitoring stations and all of them do not work simultaneously. They are capable of pointing out exceeding of safety limits for few pollutants close to their point of location, but cannot pinpoint the exact source of this pollution for sake of accountable actions. This had been the experience of Chinese authorities also, as several provincial authorities there found it difficult to prosecute the polluters despite seemingly obvious linkage. This necessitates more precise and thorough collection of data using advanced techniques.

such monitoring setups on the global scale by employing satellites to monitor how different countries are behaving with respect to their emission targets. Nationally, each country needs such technologies on a relatively smaller scale to monitor individual industrial clusters. The volume of data is huge as it is required to cover vast tracts of landmass, air and waterbodies. Also, this data about air purity is related to concentration of aerosols and polluting species as a function of distance in three planes. Many countries including China are planning to combine various monitoring techniques with big data and analytics for the sake of preventing further deterioration.[6]

Conventionally, the techniques used are based on discreet monitoring stations and on chemical reactions-based sensing. For example, a big city may have about a dozen fixed centers. These have a lot of limitations in terms of sensitivity of detection, response time, coverage area and poor inputs to modelling which is required to understand the process. The new techniques which have emerged are those based on satellite (or drone)-based monitoring and laser based remote sensing and specialised sensors. Each of these has their own advantages and disadvantages. Space-based monitoring has the advantage of pinpointing the sources of pollution grossly and eliciting the distribution pattern of pollutants and particulate matter. It can also bring out any specific geographic feature of the city/region affecting the pollution in a global manner. For example, a cement factory being close to a valley or a chemical factory facing the hilly barrier to a population center and how much this feature is influencing the pollution levels. Its inherent linkage with image processing and digital mapping techniques provides a large volume of data in a graphical manner and thus is easier to interpret. Since this gives information for various species on a real-time basis, this can serve the community in a transparent manner and ensure better compliance by polluters. The limitations of this technique are to begin with its cost. But ISRO is well known for its frugal satellite launching and the several advantages the technique offers, this

[6] Many countries have initiated efforts to monitor pollution in a way, individual countries also can be monitored in the context of their responsibilities as per Climate deal, employing satellites, Big Data and Analytics.

may not be a big limitation. The other difficulties are related to interference from non-polluting species and clouds causing complications. Also, constant observation may require a costlier geosynchronous satellite. It is encouraging that during the course of this work, the Indian government has decided to have a satellite dedicated to environment.

The second technique is that of optical remote sensing using what is called a LIDAR – laser detection and ranging, or also a laser radar in popular parlance. This is based on the principle of laser-based absorption/scattering from polluting species or aerosols. It consists of a laser system, a transmitter of laser radiation, a receiving telescope and a detector followed by data collection/presentation system. This system can be very compact and can be housed on a mobile platform like a truck, an aircraft, a helicopter, drone or a ship and there is a possibility of tuning laser colour to match and thus observe species distinctive absorption spectra of various polluting species and the beam can be rastered to probe large areas, just like a radio detection & ranging (RADAR) seeks intruding aircrafts. This has several advantages and can be quite cost effective in many situations. The laser beams can monitor otherwise inaccessible regions like the top of the chimney stack of a plant or a hilltop and track the plume and combining with the meteorological data emission rate of a particular pollutant species from a chimney can be estimated. Any non-compliance to regulatory norms can be brought to notice. The greatest advantage is high spatial (few meters) and temporal resolution (few seconds – implying almost real-time) that is achievable with this technique. Covering several kilometers with resolution of a meter is possible. Thus three or four such devices may serve a single big city. Less in number but a more skilled 0perations & maintenance (O&M) staff will be needed. Some of these features can be useful in emergency response situation, which conventional techniques are not able to offer.

Environment monitoring LIDAR has several versions which can be employed for rural or forest environment. This can facilitate in a big way the data collection related to EIA for several projects in a very short time span. Combined with global positioning system (GPS), these techniques based on space or LIDARs can be used for urban planning, industrial installations, monitoring effects of mining activity and many other functions aimed at healthy pollution free cities and rural areas. This should be part of the agenda of smart cities and villages. The essential point is that these techniques are efficient, fast, of high resolution and species selective. Once one is able to track all pollutant species, their spatial distribution and pattern of their temporal growth and linkage with a specific industrial or other activity, then it should be possible to control it in a more targeted manner rather than enforcing a blanket ban. Environment or pollution maps of a chosen geographical area can be generated at a short notice. This may influence call for recommendatory actions of various kinds. The socio-economic costs of a project can be evaluated more quantitatively and in a transparent manner. This should enable faster and right decisions accompanied by economic gains. This will improve public health and minimise expenditure on it.

Some of these systems have been developed in India but not for environment-related applications. For having dedicated systems for environment in larger numbers and for improving adaptability of these systems to mobile platforms will require some interdisciplinary work and trials on deployment. Once effectiveness is demonstrated, the number of systems can be multiplied or even the same mobile system can be transported in a truck or an aircraft. This can have huge multiplier effects in terms of its capability to handle the large volume of demand. These remote sensing devices are quite efficient to handle monitoring large volume of air (or vegetation and waterways etc.) which is constantly changing with time. This may match the dictates of 'Made in India' very well.

Lastly, the third technique is to use solid state sensors responsive to various polluting species. Many smart ways have been suggested to employ them for real-time pollution monitoring. These sensors can be combined with internet of things technology to get relevant information about position sensitive pollution in real-time. These sensors can be integrated with chosen vehicles or individual volunteers along with the potential to locate them with global positioning sensors (GPS). This way a huge amount of relevant information related to distribution of the concerned polluting species as a function of time can be generated in real-time.

All these techniques can be used in a complimentary way, depending upon the requirements. All of them are within the reach of Indian R&D establishments as the basic components and sub-systems can be built indigenously. Indian manufacturers have to take the lead by linking with these researchers. Initial action for funding and other support may have to come from state governments or even municipal bodies of big cities. Unfortunately, there are no mechanisms to make them aware of technology possibilities. Once much greater utility is demonstrated, it can catch the imagination and multiplier effects will follow. This kind of smart monitoring has to be followed by detailed analysis and subsequently well targeted control measures. Such environment monitoring need not be treated only as part of climate control commitments, but also a part of preventive national healthcare.

Components Versus Products

Any manufacturing process needs easy and economic availability of raw materials and components. Indian manufacturing units, be they in public or private sector, are largely devoted to building products and platforms, leaving the problem of basic materials and components to 'others.' This is truer when high technology products are involved. This is followed by delays in completion of projects which is attributed (one of the reasons) to what is called lack of industrial base. Many of these components have to be imported. Quite often countries have to pay heavily for these imports and wherever strategic applications are involved, there are good possibilities of them being denied. After building the products, the race for indigenisation starts. Most of the time the process of indigenisation stops at less critical components. Figure 6 shows the progress of indigenisation in a defence PSU over the years [17]. This indicates the seriousness of the situation. Some of the Indian strategic entities that are being constantly denied certain components and products have either been able to develop alternatives or have learnt to live without these. The second option is not an attractive alternative.

Besides immediate losses or difficulties, lack of easy availability of basic components and ingredients puts the innovators and designers at disadvantage. This lack of industrial base or non-availability of basic components puts many other objectives in jeopardy. Many of the FDI proposals are expected to source a certain fraction to be obtained from within the country but wherever cutting-edge technology is involved, this is not possible because of lack of industrial base. This again inhibits the process of manufacturing within the country. The new Indian defence procurement policy announced in 2016 encourages maximum indigenisation of defence-related products. But real implementation is possible only with the help of indigenous components and materials. This policy does not pay sufficient attention to the issue of indigenisation at the component level. Private players should be encouraged with 80% of development cost. Also, smaller players should be allowed to participate.

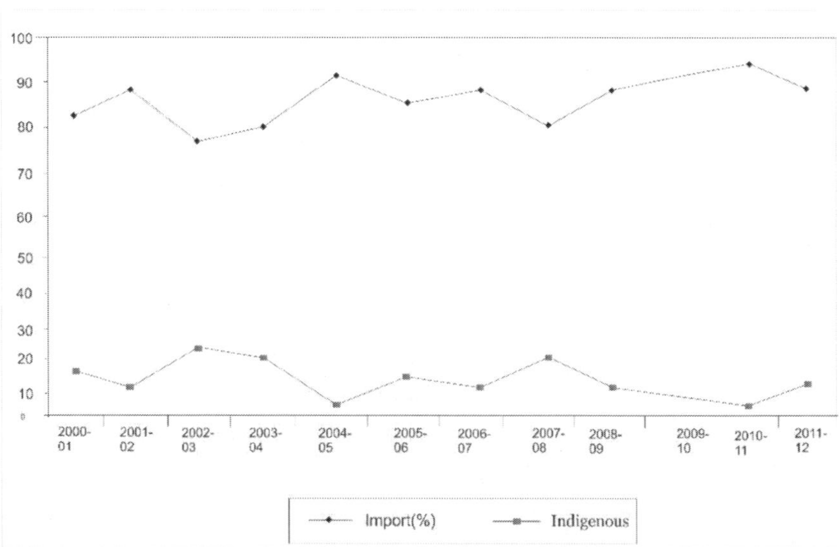

Figure 6. Shows how systems once built in a defence public sector manufacturer continue to remain dependent on imports for critical parts. Indigenous components continue to remain a small fraction in this case.

There is an estimate that in case of defence products developed and built here, more than three fourths of value addition is at the stage of system integration. A more serious issue is that some of the critical components are difficult to obtain from external sources. These components as spares affect the lifecycle costs too. Then the quality of components and sub-systems indigenised is not always satisfactory. The mindset of stressing heavily on bigger systems at the cost of smaller sub-systems and components required in much larger numbers (and value) is indicated by the fact that India has stressed on developing big systems and platforms like light combat aircraft (LCA), main battle tank (MBT) and nuclear submarine, but has not given proportionate importance to developing components and sub-systems of these big systems and much simpler products. What is true for defence-related products is equally true for most of the other sectors. Even Indian pharma industries import a substantial amount of raw ingredients from China. The Indian automobile industry has fared better in this respect but again, hi-tech components are still imported. India (both public and private sectors) has paid poor attention to building components and materials, particularly those involved in high technology. This has led to a substantial fraction of Indian manufacturing in the MSME sector being based on completely knocked down (CKD) and semi knocked down (SKD) and there is a proliferation of plain assembler-cum-manufacturers. This leads to poor value addition and affects profit margins and continued low level of skills (and thus wages) in these industries. This has to be viewed together with the fact that several MNCs are employing Indian R&D centers for their engineering design work but get manufacturing done in their home countries. If India expands and diversifies its basic industrial base of components, materials and sub-systems, then many of these MNCs will be attracted to manufacturing here. There have been recent instances of hi-tech MNCs like Apple and TESLA dithering on having manufacturing facilities in India for predominantly reasons related to local sourcing of components. Many of the budding modern industries need components and other building blocks like solar panels, liquid crystal display (LCD) screens, sensors and actuators, micro electro mechanical systems (MEMS) devices, integrated circuits (ICs) and memory chips to flourish. Precision machinery and optical and photonic components are required in industries based on technologies like robotics, IOT, smartphones and many other strategic works.

The basic components and raw materials for these new technologies, or the building blocks, should be taken more seriously. But by whom? Public sector R&D has neglected basic building blocks for too long and must look into this to facilitate manufacturing in these areas in the country. The Indian private sector could follow in the footsteps. These technology developments relate to all the sectors viz. mechanical, electrical, electronic, chemical, optical and so on. Once India is able to make good quality components at base level, then it can hope to enter the global value chains at a lower value to start with. This will also promote its strategic independence. If India inculcates a culture of developing basic level components and sub-systems, then it may not need to resort to corresponding protectionist tariff measures. These tariff measures rarely work in promoting manufacturing in India, but dedicated technology initiatives can achieve that goal.

Quality, Standards, Safety and R&D

India is more known for its *'jugaad'* (contrivances born out of necessity) technology and frugal engineering as it is influenced by demands of the domestic market where consumers have a limited paying capacity. Many of the MNCs that have adapted themselves to local market conditions understand this very well and are bringing up products to suit these conditions. A large fraction of Indian industries and even some exporters are not yet complying with international quality control standards. There is a World Bank report [8] which found correlation between standards and quality employed by an enterprise with its innovation and productivity. This report also found quality and standards poorer in India, as compared to other BRICS countries. It suggested a kind of different cultural mindset in small enterprises. For the sake of export competitiveness and many other such reasons, quality consciousness cannot be postponed for long. The recent moves by the government to promote prescription of generic drugs and the resistance of doctors to it in the absence of requisite quality control is a stark example of this. More importantly, if India has to overcome competition from other low wage manufacturing countries, then it has to carve out a niche for itself through a route based on development of reasonably priced products of high quality.

If India has to succeed in mission 'Make in India' for sectors like defence, aerospace, nuclear, railways, telecom or lifesaving medical products on a large scale, then quality and standards have to be pursued earnestly and safety and reliability norms become easy to implement. Indian automobile components became acceptable internationally first and this was one of the precursors to flourishing of its auto sector.

A large part of Indian exports includes manufactured goods and a good fraction of it comes from MSMEs. There was a survey of MSMEs of various fast developing Asian economies which pointed out that India has a much larger number of these enterprises as compared to other Asian countries, but in productivity (or export worthiness), they lag behind competing Asian economies [18]. The reasons are mostly related to quality of products exported, punctuality of delivery and lack of trust between supplier and foreign consumers. There are both technological and managerial issues at stake. Mostly, Indian suppliers are tuned to the type of products acceptable to local consumers and do not realise the type of hunger for quality and novelty which many foreign consumers have.

How many Indian enterprises work while aiming at invoking innovation and higher technology for the sake of improving quality of products? Are they able to examine domestic or international sources from where they can get upgraded technology, evaluate various sources and modify their existing techniques and technologies for the sake of improving the product quality. The World Bank report also noticed that India has a good basic infrastructure and machinery to facilitate technology absorption by small and medium enterprises, but it is not effective. This is probably because the mechanisms are governed by the bureaucratic machinery and the quality of manpower responsible for this purpose does not have the required competency level. The World Bank report suggested the same machinery to be managed by the private sector as that may improve the situation. More significantly, there is need for Indian researchers and innovators to develop simpler, economical and robust techniques for a variety of quality monitoring and control tasks which are easy to access. Preferably these should be fully home grown, the aim being total serviceability by local expertise that should also be able to cater to constantly changing demands of the industry and international level standards. In case they are costly, then they should be made available in multi-client and subsidised mode to these MSMEs. There is need to open channels of communication/forum between user industry groups and relevant R&D labs. By relevance, it is meant those who have pertinent expertise and instrumentation and not only those mandated for this purpose. Many of the Indian national labs are well equipped with the best possible analytical instrumentation, both developed locally and imported. They also have the requisite expertise to make use of them for accreditation purposes. A large number of industrial grade quality control labs can be given accreditation with the help of these national research labs. The volume of job could be big and may require some reorientation and a major extra effort on the part of these labs.

In recent times, analytical techniques have witnessed a renaissance in terms of sensitivity, selectivity, speed of sample handling and efficiency (or simplicity). Some of them have been developed indigenously also, but not translated into marketable instruments. Also, for the sake of total quality consciousness, there is need to monitor the manufacturing process right from the raw material stage to the finished product. This includes online monitoring of the process of production/fabrication, if possible. This also calls for attention to the requirement of making the technique user friendly or/and training adequate number of skilled operators/technicians for the sake of reaping the benefits of multiplier effects.[7]

---

[7] For improving preventive healthcare of its citizens, some of the commonly used edible products like water, milk, oils etc. need to be monitored for their goodness by very economic means and these should be accessible to large fraction of poor population. While scientific researchers need to develop appropriate techniques for this purpose regulatory authorities need to tap them suitably. Researcher has to make these techniques operator/user friendly.

Such instruments are available in the international market. These are quite costly. There can be several techniques based on optical spectroscopy, nuclear spectroscopy and mass spectrometry which could fall in this category. Adaptation of many of these techniques presently employed in research labs by manufacturing industries is a critical missing link. The reason being either costly or difficult to operate or bulky and difficult to transport. Indian researchers need to overcome these hurdles and find universally acceptable solutions.

A major example for quality monitoring is that of Indian herbal medicines. Herbal medicines have been important ingredients in human life for generations and have recently acquired increasing importance due to their supposedly harmless nature (lack of side effects) and effectiveness. More significantly, these medicines still serve a large fraction of Asian population which is poor and which lives in remote areas. The common problems, of utmost interest, affecting their safe usage, involve use of fake or similar but wrong herbs, their contamination and lack of standardisation. The traditional Indian medicinal systems such as Ayurveda and *Sidd*ha also recommend use of metals in small amounts (nutrient and heavy metals) in herbal medicinal preparation (called bhasma) for different ailments. An excess or deficit of such metals and their unprincipled use may cause unknown repercussions. Therefore, a rapid, simple, and accurate analytical method for identification of herbal plants, including distinguishing between similar species growing in different geographical locations, and trace element detection, is highly desirable. The problem is to make these medicines acceptable to a wider base of international customers the way these are accepted by Indian patients. For Indian patients, it is a matter of trust developed over centuries (even Indian patients could benefit from these improvements in quality monitoring and control), which is not true in the case of foreign patients and their regulatory authorities. Despite the proven potential of spectroscopic techniques for such analytical/ quantitative measurement, these have not been systematically applied to Indian herbal and Ayurvedic medicines. Several Indian universities have worked on these problem areas and efforts should be launched to collate these results and establish a more efficient method of standardisation based on intercomparison of results from various laboratories. This will augment the scientific basis and safer manufacturing and use of such traditional medicines and will encourage an alternative medicine system as well as promoting exports of these medicines. Researchers have to take care that in the process, the cost of the medicine does not go up very much[8]. There could be easy methods for several types of food products or even water which may require quality monitoring and control. These are a few example of R&D in useful areas with a lot of potential for public good. There are a large number of such cases where R&D has been done to a mature level, but the results have been dismal. A major reason behind this is the inherent problems with the R&D system which has poor linkages, lack of trust, lack of motivation etc. This segment of R&D has to address these issues for extending the customer base and making it more acceptable to international regulatory bodies. Similar to herbal medicines are cases of several food products (milk & water) of common use and several industrial products and processes of all kinds which need to be upgraded in terms of quality and safety. If technologies are easily available, regulators may find it easier to enforce relevant legal requirements.

As an illustration, only quality/purity related to chemicals or pharmaceuticals have been touched upon here. Similarly, various techniques have emerged with applications geared to precision mechanical or electrical measurements and better standardisation in mechanical and electrical engineering. Extensive related R&D is being done in various national labs and it needs to be tapped, expanded in scope while establishing linkages to manufacturing. In addition to quality, related problems are those of standards and safety. Extra load on some regulatory agencies, corruption and poorly trained staff (not abreast of latest technologies), lack of genuine interest in such R&D areas has hurt efforts related to these issues. Good R&D existing in these areas can be translated to useful deployment. These challenges are related to various disciplines of

[8] New Board on Ayurvedic and indigenous medical systems has been activated. But this will see genuine progress when it joins up various analytical R&D laboratories to harness modern technologies for economic and efficient quality monitoring and control of these medicines.

physics, chemistry, biology, metallurgy and several other types of engineering. The motto for Indian researchers should be to develop quality monitoring and control techniques which are economic and easily accessible to all. The aim should be to make Indian products of all kinds of high quality and yet not very costly.

Managing Disruptive Technologies

Some of the new technologies on the horizon are suspected by many to be disruptive. The most serious threat of this disruption could be to developing countries, particularly the ones with large and unskilled population, which may have to compete with machines. India, which was expecting to harvest demographic dividends, may be the loser if it fails to respond properly to these new technologies like artificial intelligence, robotics, additive manufacturing, internet of things, nanotechnology etc. The thrust on development of these technologies cannot be influenced or resisted by anybody. This may call for a well-crafted adaptive and evolutionary mode of response. While India should not shirk from participation in development of these technologies, it must exercise caution and wisdom in deployment and application of these technologies. This is where the role of innovators, researchers and technologists comes into play. Many western countries aim to improve their manufacturing capabilities resulting in what is called industrial revolution 4.0. One of the major goals of this is to reduce cost of production and reduce dependence on labour force from developing countries. Development of software for novel applications of artificial intelligence, robots, drones and R&D for development of sensors and actuators associated with deployment of robots form the basis of growth of such technologies and could stimulate a host of new applications. India has reasonable potential in many of these areas, but its researchers have to be imaginative in finding novel applications rooted to its domestic techno-socio-economic requirements.

Advanced manufacturing techniques have different connotations for countries with different technological, socio-economic and cultural backgrounds. In India, one can say the same to be true even for its different constituent states. Internationally, manufacturing technologies are taking a new avatar in recent times because of various reasons both on the side of supply as well as demand. On the demand side, these factors are like appearance of new materials on the industrial scenario, handling of materials at micrometer and nanometer level, so called composite and smart materials on the scene and demands from applications in bio-technology, ICT and microelectronics industry. Integration of digital-based technologies into manufacturing continues to change the complexion of processes of manufacturing. Some of these technologies which may sound futuristic now may rapidly burst into the commercial scene. In case of a hesitant response, India may miss again an opportunity to make a leap.

Among the fabrication/manufacturing techniques a revolution of sorts is taking shape in the form of advent of additive manufacturing. This technique is in contrast to conventional manufacturing which involves removal of material in a pre-designed way for generating a desired shape, as if a stone is cut by a sculptor by subtracting material from original piece. On the other hand, in case of additive manufacturing, a material is given shape by adding layers of chosen materials on base as per design requirements. Saving in terms of material consumption is inherent to the technique. In additive manufacturing, with the help of new tools of "directed energy", sprayed powders of constituent material are first molten (in case of metals) and then solidified into the desired shape by means of layer by layer addition. These "directed energy" tools could be in the form of laser beams or electron beams. For a fabricator, this implies moving away from the mode of assembly line production to customised single unit production. If the machinery can be made simple, even a user (industrial or domestic) can himself produce an item to his individual satisfaction. The skills required for additive manufacturing are totally different than that for subtractive manufacturing.

National Aeronautics & Space Administration (NASA) engineers have produced a part of a rocket engine using this technology [19]. This part, a combustion chamber made of copper, was a difficult test case with more than 8000 layers of molten copper powder and was prepared in about ten days which is significantly less than conventional subtractive manufacturing. NASA engineers see it as the beginning of a journey of a technique to be useful to inter-planetary explorers in future for fabrication as well as repair of spacecraft components while in space travel. Similarly, Chinese scientists and engineers used this technique and produced the front windshield frame of their recently launched transport aircraft. A single piece came for 500000 dollars and was to be imported and to be delivered in two years. This was produced indigenously in China using additive manufacturing (AM) technique in 5 days at one tenth the cost [20].

What are the implications of this technology for a developing country like India? The first demand on any new manufacturing technique is that any such large-scale deployment in manufacturing activities should not affect employment. Economics consistent with quality, simplicity and amenability too low to medium skill workers are other demands. Indian innovators have to exercise their imagination for the most optimum deployment of such new technologies. So, more specialised applications should be of interest. In sectors like aerospace, defence and nuclear, or railways there are frequent demands on low volume of customised fabrication (or repair) or in situ manufacture/fabrication of major equipment or spare parts, involving many a time materials which are precious or where replacement of parts is costly. Chinese reports suggest [20] a cost reduction by factor of ten or more, large improvement in delivery period and up to 90% saving on raw material. There are several other sectors which could benefit from this like the construction sector for rapid affordable low-cost housing, customised human body parts making use of plastic waste for recycled products, rapid prototyping in industrial applications, disaster relief operations for speeding up urgently needed specific deliveries at remote locations, models for science education, architecture and entertainment applications etc. [21]. Many of these specific deployments of this technique may not have got international attention, but has a great scope in developing countries like India. Progress in such technologies needs to be dispersed widely for the sake of multiplier effects.

Some of these applications have been achieved in India in a few isolated cases, but real benefits will accrue from the massification of decentralised deployment. This technique is amenable more to dispersed deployment rather than large scale assembly line production as mentioned earlier. There are a few startups and institutes that are working on localised problems related to 3 D printing. A Bengaluru-based startup has made body parts like human liver using this technique. Another innovative startup has developed a hybridised version employing both additive and subtractive machining in a single low-cost equipment. For facilitating large scale harness of this technology within the country for various applications, the Indian R&D sector may have to develop low cost equipment for this purpose and train the required workforce to run and maintain that equipment. At this stage of technology development, it may be difficult to predict the course that various applications of this technique may take. For industries and young startups, development of rapid protyping centers for manufacturing could be a boon. A good number of such centers, well distributed geographically, will facilitate young manufacturing startups to play with innovative fabrication. Much more is dependent on imagination of designers, innovators and engineers. Widespread availability of these machines and experiments with a larger range of materials is crucial for MSMEs. Since the Indian government has declared a major new initiative titled 'Digital India,' this will intimately link up the nation to these technologies. Broadband access and 5G telecom in remote areas with the help of BHARATNET infrastructure could offer a lot of distributed manufacturing possibilities.

Challenges for Researchers

Manufacturing technologies in India can be broadly divided into three categories viz. traditional, strategic and frontline. Among these, innovation and R&D culture has matured fairly in the strategic sector, but presently it contributes only a small fraction to manufacturing in India, the defence sector being a case of specialised and urgent requirements. It is evident that there is a good amount of role for scientific research and technology development in bringing up mission 'Made in India' in the context of several traditional sectors. There is plenty of knowledge base/technology that can be transferred from its R&D entities to manufacturers, with or without invoking patents. Noticeably, the existing R&D results also need to get deployed fast for manufacturing applications. The choice and mode of deployment of technologies has to be made carefully to suit Indian socio-economic requirements and Indian researchers and technology developers need to be more original. The R&D efforts in areas of disruptive futuristic (or frontline) technologies may facilitate better assimilation and adoption of such technology by entrepreneurs in future. These efforts have a good potential to provide increased quality employment. Thematic stresses have to be encouraged in new areas like for example environment, basic components, cyclic economy, 3D printing and other advanced manufacturing technologies. These themes of R&D facilitate removal of barriers to manufacturing and offer fresh opportunities for enterprise. It is not often appreciated that R&D and action on environment for India is not purely for the sake of international climate-related commitments but for the sake of the health and well-being of its own citizens. The purpose of scientific research on total quality monitoring and control is multifold. While the Chinese mostly sell low quality products at incredibly low prices and western countries sell high quality products at high prices, it should be the endeavour of Indian researchers to develop a niche of high quality, low price products. Several such goals could stimulate many R&D areas.

Innovation can cut across institutional boundaries and so it is not necessary that one falls back (or is strict about this) for technology in a particular area on an institute meant for that subject. This implies that technology flow should not be mandate limited and cross-sectoral technology contributions (and collaborations) need to be encouraged. The new-found enthusiasm for fresh enterprises or tech-based startups, confined so far mostly to softer sectors like IT-based ones, has to be exploited for linking research to manufacturing. Startups can play a significant role for this purpose, generating employment. Spinning off some of these enterprises/startups from national labs also needs to be encouraged. STEM higher education and generating appropriate skills as inputs to these startups will be an integral part of this scheme of things. Basic sciences have a big role in improving human resources. All these demands call for a major realignment of priorities for various research institutes and manufacturers. This calls for reforming, reorienting as well as enhancing support for researchers and thence to manufacturers. A well-designed symbiotic system of R&D efforts for industrial manufacturing in the country needs to be organised. Presently, the government R&D is responsible for nucleating most of the strategic, advanced and core technologies. Special efforts and environment generation is needed to go beyond this and rope in private sector by initially starting with joint ventures.

These are all big demands and the main battlefront is on attitudes and mindsets. There is need for India to examine in detail status, mandate and interaction mechanisms of its research labs — between themselves and with manufacturers and institutes of STEM HE subjects. How are Indian innovators, scientists and technologists sensitised to the issues? In what way national policies on related subjects influencing the progress on the innovation front? While utopian equality can be a goal of economists and political class, development of technologies for inclusive, accessible and sustainable growth for all is what scientists and technologists in India can aspire for. This may require a major cultural shift. These are not the goals which scientists and technologists usually work for. Some of these issues are discussed in the next chapter.

Chapter 4

INDIAN RESEARCH AND DEVELOPMENT LABS: EVOLUTION AND IMPACT

INTRODUCTION

In India, a large fraction of population is not yet literate and less than a quarter of it opts for tertiary education. What constitutes a modern laboratory is still a mystery to the general public. A good fraction of educated population in most developing countries is exposed to science through experiences at school level science laboratories. In India, for example, a large number of these labs are quite antiquated. People's interaction with science is limited and is mostly restricted to their visits to science museums and centres in big cities. On top of that, the quality of science teachers in majority of schools in most developing countries is deplorable. Collectively, the impressions in the minds of public at large about science, technology and the role they play, or can play, in economic progress or other national affairs, is far from reality. A large fraction of major decisions taken by the governments (both state and central) have some relationship with technology. There is a strong need to inform and educate the public about some of these aspects of science research. The public at large presumes that a research lab produces novel things. Only a few understand that research labs mainly build upon concepts and ideas and either work for pushing frontiers of knowledge or take ideas to the level of a new product or process and are not supposed to be involved in scaled up production. Educating the public about science is required on several fronts. There is greater need for popularisation of science, teaching the ‘teachers’, equipping the village level school laboratories (of course after proverbial ‘blackboards’!), having more scientists or technocrats in policy making and sensitising the policymakers about the benefits of being in the frontline of international science and technology.

India has a large network of research labs, besides several institutes of higher education engaged in R&D. Some of the Indian national laboratories are those working in the areas of space, defence, atomic energy, agriculture and biotechnology. A battery of scientists is also engaged in Council of Scientific & Industrial Research (CSIR) labs. Today, national laboratories are conglomerates of several laboratories working in multiple disciplines. What is the structure and mandate of these national labs and how they have evolved and kept pace with advances in science and technology in various countries? How these labs have linked themselves to the society in general and industries in particular? Some of these questions with specific attention on Indian labs are discussed in the following work, with a brief introduction to national laboratories in the world.

National Labs, Technology and Manufacturing

Historically, natural sciences have grown mostly as a result of individual enterprise of scientists with modest support from governments or private players in the form of philanthropy. It was industrial revolution in the western world in the 19th century that changed the scene as science and technology development started overlapping. Research started being viewed as basic, applied and technology development. The last one did impact economic progress of nations. Many of the financiers jumped on the bandwagon and started investing in research and development with the hope of reaping profits. This paved the way for emergence of industrial houses having their own R&D labs. The period of discussion is close to the end of the 19th century and beginning of the 20th century. This was the time when telephone, wireless communication, aircraft, and electricity had started affecting the lives of ordinary citizens in the western world. This was the time when patent wars began. There is the famous case of Marconi and Tesla about wireless communication and those of Westinghouse, Edison and Tesla about electrical lighting, distribution and AC versus DC transmission. It was Edison who combined large scale teamwork for invention and the principle of mass production for the first time. That was the beginning of industrial research. This required large finances, unlike small grants required for scientific labs earlier. This was also the beginning of linkage of technology development with large finance companies like Morgan and Drexler. The same company financed Tesla, the later day competitor of Edison. This could be called forerunner of present day venture funding.

Another major impact on the evolution of big labs was that of World Wars (I and II). For example, Boeing Airplane Company was established in 1917 and soon got big orders from the US Navy. Their profits in wartime efforts were ploughed back for research and development. It also called for major readjustments on part of these R&D-cum-industrial companies to peacetime conditions. This contributed to the development of civilian aircraft. Another such aircraft company to have similarly benefited from the war t was Hughes Aircraft Company. A parallel growth was that of big pharmaceutical companies. The war required big stocks of medicines for soldiers. Some of these small (at that time) pharmaceutical companies like Pfizer which were involved in small time R&D got big orders leading to a major growth of R&D in pharma industry. It was also the period when industrial companies merged together with R&D labs for the sake of mutual benefit. An example of this is in the Western Electric Research Laboratory and the engineering department of American telegraph and telephone company coming together to form the phenomenal Bell Labs. The management of Bell Labs encouraged some basic researchers for the sake of long term benefits to the company and this yielded them rich dividends as over a period of time, this lab claimed seven

Nobel Prizes and contributions to major researches bordering applied sciences and technology.

As these major industrial labs were growing in the beginning of the 20th century, fundamental sciences, particularly physics and chemistry, were making great progress and most of the research was carried out by individual efforts. This research was, for example, Albert Einstein on photoelectric effect and the theory of relativity, Max Planck on the theory of radiation, Niels Bohr on structure of atom, Heisenberg and Schrodinger on quantum mechanics, Marie Curie on radioactivity, Alexander Fleming on penicillin and many others. Most of these efforts were centred in either Europe or America. On the other hand, there was slow but steady progress towards mechanisms of collaborations between research labs and industries, with technology transfer as a consequence. The United States is an excellent example related to evolution of links between university research and industries. In the United States, the government patronage to private research and technology development by private sector was governed by an old 19th century law called Morrill Act. As per this, the only way researchers could benefit common public was by means of research publications. Researchers in those times considered private grants as a 'necessary evil' and accepting royalties was considered to be a taboo. Slowly, private funding sources (corporations) started funding projects with commercial worth and demanding the right of first refusal and alternately exclusive rights. The University of Wisconsin was one of the first universities to have patents filed on the basis of research by its faculty in 1924. This was followed by the great depression and consequent financial pressures on universities and also technology demands of the government. This catalysed the process of technology transfer by academics for public good.

A major shift in the situation can be perceived when nuclear fission was discovered. Many scientists could foresee what this discovery could unleash. This happened around the onset of World War II. There was great fear in the minds of many scientists that the Axis powers could exploit it for its destructive potential with disastrous consequences for the world. This led some of the scientists like Leo Szilard to persuade Albert Einstein to write to then American President to take the initiative before it is too late. Einstein had great moral authority and the President of USA could be persuaded to launch the greatest initiative in the history of science and technology linked to industries. This was the beginning of the constitution of a major American national lab devoted predominantly to national security and related wartime efforts. This was founded in 1942 as an offshoot of University of California as it was realised the work on experiments related to building nuclear weapons could not be done in a single university and a multidisciplinary and multi-national team of researchers was required to work on such a programme. This laboratory handled a major classified project called 'Manhattan Project' for testing and building nuclear devices. Having accomplished its wartime goals, it adapted to non-war research after the war ended. This laboratory could be called the first truly major national laboratory (Los Alamos Laboratory). Soon after the war, another competitor lab presently called Lawrence Livermore Laboratory, was established. It was with major persuasion from E. O. Lawrence (father of particle accelerators) and Edward Teller (father of thermonuclear device) that this laboratory was setup and was an offshoot of Lawrence Berkley Laboratory. Lawrence was a big advocate of "big science" tackling challenging projects by deploying physicists, engineers, chemists working together in multidisciplinary teams. This set the trend for large sized national laboratories working on a broad mandate. Another major national lab was setup in USA in 1958. NASA was originally meant for civilian and peaceful applications of space exploration. But soon America got embroiled in the Cold War and the race for prestige and security began, especially after the Soviet Union launched its first Sputnik in space. In 1961, President Kennedy announced the launch of a major national effort to send man to moon and bring him back successfully. This was achieved by 1969 with the help of Apollo mission. It was one of the costliest research and development projects ever. It costed 205 billion dollars, as compared to about 26 billion dollars for Manhattan project. All these permeated some sort of 'halo' around many national labs the world over. Similar big national laboratories sprung up in many other countries over a period of time. These developments were taking place in

America, Europe and Soviet Union and all of them were progressing in parallel. While the Soviet Union was in competition, European countries were independent with limited collaboration with Americans.

These projects by the US government, besides their scientific and technological significance, provided a major stimulus to American manufacturing. However, within a few years, there was a perceptible decline in manufacturing and the economy shifted towards services. The American government made a major effort to boost manufacturing and succeeded well. In 1980, it passed a law called Bayh Dole Act, which changed the scenario of patenting and technology transfer. This law served the purpose of clarifying situations related to ownership of proprietary technologies and their transfer. The next major step was Federal Technology Transfer Act, passed in 1986. Government laboratories were allowed to agree in advance to license any invention made during collaboration with private entities by the federal employee to the collaborating party. Some of these measures gave a major boost to technology transfer and its commercial exploitation for manufacturing and job generation in the hi-tech sector. These clarifications about incentives to researchers could be considered a major boost to linking university research to manufacturing.

It is quite evident that for research to be linked to manufacturing, industries in a big way required a driving force like a world war or "cold war," or even a race with Soviets for national prestige. But besides the driving force, a favourable legal framework taking care of all players involved did play a crucial role. In most of these cases, it boosted development of science and technology and provided employment to a large fraction of population. In developing countries, such driving forces for scientific research and technology development could be poverty, hunger, malnutrition, diseases, poor connectivity, and supply of energy. How successful these countries are and to what extent science has helped them?

MAJOR INDIAN RESEARCH LABS

Lack of political independence did have a major influence on lack of growth of science in India. There were a few gifted individuals who contributed to science and won laurels. Among them being J. C. Bose for wireless communication, C. V. Raman for light scattering, S. N. Bose for statistical physics, M. N. Saha for astrophysics and so on. But there was another set of individuals who thought in terms of 'national science' and its development. Among these were H. J. Bhabha, S. S. Bhatnagar, P. C. Mahalnobis and some others. Before independence, there were clear restrictions or lack of support from the British government of the time for scientific or industrial research. This is clearly indicated in the comments of then viceroy, Lord Willingdon, to a note prepared in 1934 for his approval for establishing a council for scientific and industrial research. Lord Willingdon rebuffed the move commenting tersely, "it is not necessary." But persistence of dedicated personalities like Ramaswamy Mudaliar (member executive council of viceroy) and Dr. Shanti Swarup Bhatnagar that led to the creation of CSIR in 1942 and the first five national laboratories were established by 1950. The political establishment in independent India proved to be a big help in all these pioneering efforts. CSIR is now one of the biggest industrial research organisations in the country with the largest number of patents to its credit.

There is a clear contrast to agricultural research in the country. The British government was too eager to promote agricultural research in the country. On the recommendations of the Royal Commission on Agriculture, Imperial Council of Agricultural Research was established in 1929. This was not to help the famine-affected populace but to predominantly encourage research on improved varieties of cotton. Textile tycoons of Manchester kept demanding better yield and variety. This discriminatory attitude to research on raw material/commodity compared to that on manufactured goods bore the stamp of a colonial mindset.

Indian universities and other institutions of higher education and research in general had been quite lethargic in developing an attitude and mindset of contributing to technology advancement related to industrial growth. This was despite the fact that Indian Institutes of Technology (IIT's) were modelled on the lines of Massachusetts Institute of Technology (MIT). In fact, this modelling worked contrary to expectations and a large number of students from these institutions felt as if 'trapped on an island of technology excellence' with a strong urge in a large number of them to migrate and flock with their own 'technology cousins' abroad. This was the price India paid for focusing on a few islands of excellence and ignoring mass institutions/universities. Many realised that unless a critical mass of talent is formed internally, it will be difficult to retain talent and encourage technology growth. Once such an ecosystem is generated with R&D directions better aligned to industries, there will be greater possibilities of industry-academia interactions.

Indian national R&D labs on the other hand received a very favourable treatment and were organised around very specific mandates. There were a few very gifted scientists who worked tirelessly to bring India on the global map of R&D. But its linkage to manufacturing was not so much on their minds, despite the formation of CSIR. This had reasons much beyond the immediate brief, which will be explained later. A brief description of some of the major Indian national labs and their present role as well as potential in promoting industrial growth is discussed in the following.

Industrial Research, Atomic Energy and Space Research

Dr. Shanti Swarup Bhatnagar led pioneering efforts to initiate scientific research related to industries. He worked hard to set up a chain of laboratories working on various themes with linkages to industries of contemporary interest. Parallel to the efforts of Dr. Bhatnagar were those of Dr. Homi J Bhabha in trying to set up a laboratory for fundamental research on one hand and nuclear science and technology establishment on the other. He pointed out the justification for theme-based national research laboratories in his last lecture delivered by invitation of International Council of Scientific Unions in January 1966 [22]. He said, "The universities are, however, autonomous organisations, jealous of outside interference, and the process must necessarily be slow and time consuming. It is probably for this reason that it has been considered expedient in many countries to set up national research laboratories and other scientific organisations for specific subjects." After huge efforts, he was able to establish what is presently called Tata Institute of Fundamental Research (TIFR) with the help of Sir Dorabji Tata Trust and the Bombay government. This institute did have some links with Bombay University through some professors of the university.

A third major effort in India was in the foundation of Atomic Energy Establishment (AEE). This was a unique experiment on part of a great visionary, Dr Homi Bhabha. His aim was to have a self-reliant nuclear science and technology centre. He had clear goals set for developing know-how indigenously and on increasing the number of people who were able to tackle major technology tasks ahead. The generation of self-confidence and the ability to engineer and execute industrial projects without foreign technical assistance was another objective. He had full faith in the capabilities of young Indian scientists. He was paid back in ample measure by the achievements of what later became the Department of Atomic Energy (DAE). Bhabha himself pointed out to the contrast in the two approaches — growth of CSIR and DAE. Bhabha worked hard and exploited his excellent rapport with then Prime Minister Jawaharlal Nehru to remove a plethora of administrative irritants and hindrances in setting up a huge atomic energy establishment. While CSIR went by the normal administrative rule book, AEE and TIFR could break those shackles and race ahead. Two biggest gifts which Dr. Bhabha can be considered to have given to the nation were a novel (by Indian standards) and smoothly working organisational structure of the scientific institutions he set up and the level of self-confidence he generated among his scientific staff. He believed this confidence to percolate down to interacting industries, universities and administration over a period of time. He used to confide to his colleagues that the introduction of atomic energy in India in 1950 was not purely for the sake of atomic power but for the introduction and development of frontier technologies in the country. This organisation continues to work for nuclear power and pursuing research related to strategic programmes. But percolation of this culture was quite limited, as the nuclear industry remained a small fraction of total manufacturing in India.

The Indian space programme and related national labs were born as a result of efforts of Homi Bhabha, Vikram Sarabhai and Satish Dhawan. They had the sagacity of working together as a team and persuading the Government of India to launch major programmes on space-related research. As a result of major expansion of this programme, a totally independent department, ISRO, was created. This organisation had its genetic makeup similar to that of DAE. There large part of senior scientific staff of ISRO owed their training to DAE. With less of bureaucratic chains, this organisation too flourished with several successes. This organisation also has several laboratories working on different areas but with overarching theme of space research. ISRO has reoriented itself by focusing on various missions of public interest like those related to environment, navigation, telemedicine, remote sensing, cartography and educational applications. It has contributed significantly to military applications as well.

While CSIR is the research body considered to be promoting industrial and manufacturing activity, DAE and ISRO have only indirect links to manufacturing. DAE has so far been responsible for building and operating nuclear power plants. All associated R&D and human resource development is also their responsibility. Presently, about 3% of power produced in the country is of nuclear origin for which NPCIL is responsible. Similarly, Department of Space programmes have a huge potential for scaling up and can consequently impact manufacturing activities in the country. Presently, these contribute a very small fraction to the Indian manufacturing sector as most of related manufacturing requires high skills for which Indian industries are not yet prepared. CSIR labs have wide spectrum of science & technology (S&T) areas with influence on manufacturing. CSIR labs are responsible for the largest number of patents of an R&D lab in the country. Only a small fraction of these get converted to industrial manufacturing. CSIR is one of the few scientific bodies which is evolving the Indian Patent Rights (IPR) policy on its own and is interacting with several Indian as well as global industries. Their best-known contributions are those related to growth of Indian pharmaceutical manufacturing. Besides scientific research and technology development, all these departments have good potential of contributing to human resource development. For various reasons, the impact of this R&D organisation on manufacturing has been limited.

Defence Research

One of the largest R&D organisations in India today is DRDO. This organisation covers wide interactions with production agencies and autonomous user groups. Its work areas deal with mainly applied sciences, technology development and manufacturing. This organisation had a humble beginning with what was then Defence Science Organisation with limited mandate related to R&D in the areas of physics, chemistry and mathematics which bordered defence applications. There were independent organisations of technical development establishment of the Indian Army and the Directorate of Technical Development and Production which were merged with it to form DRDO. This was as per the recommendations of a committee headed by P. M. S. Blackett, a British Scientist. Blackett had a very good rapport with the first Indian Prime Minister, Pandit Nehru. Blackett's recommendations suggested a short-term plan for indigenous development and manufacture of what he called 'non-competitive weapons' and a long-term plan for development of relatively complex weapons. There was no suggestion to take up advanced weapon systems. This was as per expectation, as any foreign scientist, howsoever big, has limitations in appreciating the true potential or ambitions of a nation. The wars of sixties, which India fought with China and Pakistan, influenced defence research establishment. The R&D budget was augmented and some restructuring was done to separate R&D from production. There are several ordnance factories and public sector units associated with DRDO. One of the major successes of defense research today has been development of several missiles. However, Indian defence users have huge demands from these organisations. Presently, this organisation has more than fifty laboratories covering a vast range of science and technology research sponsored by the armed forces. Defence research has major implications for the mission 'Make in India' as a huge internal demand exists and this sector is becoming a big burden on Indian imports bill. Recently, the government has opened up this sector to private players and even FDI has been increased to 49%. Some of the erstwhile classified projects are being passed on for development to private sector consortia or even consortia of IIT departments. Prior to this, DRDO has been solely depending on several DPSUs and ordnance factories for their manufacturing activities. There are huge possibilities of domestic manufacturing based on R&D and technologies developed by DRDO. To keep pace with the demands of its users, it may have to deliver quality products in a timely manner and outsource more of its scientific research and technology development to suitably chosen partners from STEM HE institutions and consortia of private industries.

Agricultural Research and Manufacturing

As mentioned earlier, on the recommendations of the Royal Commission on Agriculture, the Imperial Council of Agriculture was set up in India in 1929. Before this, following famine of 1901, Lord Curzon, in 1905, got established the Agricultural Research Institute in Pusa, Bihar. Most of the activities though were related to education and training and research could not be carried out for want of finance and competent research personnel. The council was expected to govern and coordinate research activities in the provinces. India is a vast land with several different climates and soils. Accordingly, several commodity committees were working with an aim to specifically research on each one of them. For example, the cotton committee was established in 1921. These committees were semi-autonomous and financed partly by the government and partly by taxes collected from export of these commodities. The main thrust of the British government was on cotton, tea, silk and indigo.

In March 1947, the council was renamed as the Indian Council of Agricultural Research. The first serious effort on agricultural research was with the initiation of projects for intensification of regional research on cotton, oilseeds, and millets (1961). As many as 70 projects for different crops were launched. By 1965, most of the earlier commodity committees were abolished. Agro-research got major recognition with the help from US department of agriculture. Some of the major milestones that were achieved included the launch of hybrid maize in collaboration with Rockefeller Foundation, establishment of first agriculture university in UP and agriculture colleges in different states. In late sixties, with stress on minor irrigation and hybrid varieties, the country ushered in the Green Revolution. In 1972, ICAR director general's powers were enhanced and he became secretary to the Government of India. Under agriculture scientist Recruitment Board, agriculture research service for recruitment of scientific personnel was initiated. Depending upon local ecology, the entire country was divided into several zones. ICAR was not only expected to manage research, but also act as main clearing house for a variety of information to farmers and agricultural producers.

India and its vast populace are heavily dependent on agriculture for food and livelihood. Also, even though the main responsibilities were related to agriculture and veterinary research, there were implications for assuring fair price to farmers and regulate food supplies. Accordingly, Agriculture Price Commission and Food Corporation of India were established. The fact is that there are many more aspects which influence agricultural produce and its reach to the public. Agriculture research has been doing well to serve the nation but with increasing population, the burden to increase the efficiency, quality and productivity provides major challenges. There are many research gaps which have been identified by the department. This department has set a good example of how fruits of research were translated quickly for actual deployment on land. India turned into a food surplus country as a result. There are still some areas of concern like pulses and oilseeds.

Agriculture related R&D has plenty of areas of multi-disciplinary research which depend on other departments and integrated planning is expected to contribute tremendously to the progress. These could be prevention of wastage, monitoring biotic stresses on vegetation by space-based or land-based remote sensing techniques, food preservation, storage and transportation, use of bagasse or biomass for co-production of energy , contributing to fuel production (or substitution of petrol products), herbal pesticides and efficient monitoring of crop inputs and food quality.

Although in the parlance of economists, agriculture may be treated as independent of manufacturing, but it has a lot of linkages with manufacturing activities on agricultural implements and machinery, large sized irrigation projects, drip irrigation techniques, use of renewable energy for water pumping by replacing the conventional method, food processing, generation of renewable energy using biomass, fertiliser production and controlling pollution (of bioenergy generation) by proper disposal of agricultural bio wastes. Many of these R&D areas can influence agriculture-related manufacturing industries and the 'Make in India' programme. These areas cover a vast landscape which requires collaboration with several other R&D institutions.

Pharmaceutical and Biotech Research

The industrial policy resolution of 1948 viewed knowledge and expertise of foreign companies as of significance to industrialisation of the country. Drugs and pharmaceuticals were put in the category requiring a high degree of technical skills. But soon it was realised that these foreign companies were behaving like trading agencies — importing drugs and marketing them in the country and not contributing to increasing the competence base. Accordingly, the industrial policy resolution of 1956 reframed the policy in this regard. Industries were categorised in such a way that drugs and pharmaceuticals were placed in Schedule B, giving public sector a leading role in development of this sector with some support from private players. In 1977, this was reemphasised with an aim of taking help from ancillary and cottage industries. It was preferred to outright purchase of best available technology and adapting it for the country's needs. In pursuit of this, five public sector companies were established. Out of these, two played a very important role in contributing to a robust private sector. These were Hindustan Antibiotics and Indian Drugs & Pharmaceuticals. Their efforts were supplemented with help from WHO, United Nations International Children Emergency Fund (UNICEF) and Russia. Their healthy growth gave the confidence that the bulk of drugs could be built in India. It also prompted the Indian education system to respond suitably and provide suitable skilled manpower.

Also, some of the CSIR labs, like Central Drugs Research Institute, Indian Institute of Chemical Technology and National Chemical Lab were developing processes for development of several drugs and facilitated private sector industries to scale up these processes to industrial level. Some of these companies mastered the art of reverse engineering and the lag time for new pharma products from global markets to Indian market was reduced to about five years. This was thought possible because of the patent law of 1970 for processes instead of products. Indian companies had to establish only bioequivalence of drugs. In 1991, the industrial policy was changed to give private sector a lead role and public sector was expected to support. This step might have helped private industries, but it had a deleterious effect on public sector R&D growth.

In an investigative study, Balance [23] found that Indian companies were competent in handling reverse engineering but were not capable to handle all stages of drug development. This aspect was ignored while allowing 100% FDI in pharma sector. Also, in compliance to trade related intellectual property rights (TRIPS) under World Trade Organization (WTO) obligations, product patent regime had to be enforced. Over a period of time, the consequences have been that most MNCs are setting up shop here mainly for the sake of conducting clinical trials with the help of Indian collaborators. They are interested mostly in brownfield investments. Their R&D intensity within the country is lower than that of even Indian public sector institutes. In the process, incomes of local pharma companies might have increased as a result of these collaborations, but the overall indigenous competence base could suffer a blow over a period of time. This may also result in neglect of R&D work on diseases of interest to

India. Any MNC that ventures into a new drug research may require investment to the tune of one billion dollars or more, and will like to make proportionate profit. It is imperative that India envisions a plan for the poor people of Asia and Africa keeping in mind the commonality of many diseases. Indian public sector R&D institutions need a grand revival. Indian private sector R&D labs, built quite assiduously, have joined the global supply chains and pubic private partnership (PPP) policies are not yielding so much. The Indian brand in the newly discovered drug domain is yet to be respected by doctors and drug controllers worldwide. Public research under the auspices of CSIR and Department of Biotechnology, like New Millennium Indian Technology Leadership Initiative and Small Business Innovation Research Initiative need to be continued for the sake of more active collaboration between public and private sector R&D labs in the pharma sector. This will help in reviving the former and bringing back the latter into national R&D frontiers. At the same time, global public health initiative has to be pursued for promoting R&D in several fund starved R&D areas. Many of these problems will be of interest to several Asian and African countries. Such research initiatives can ultimately be a big boon to manufacturing in this sector in India.

Renewable Energy R&D

The term renewable energy gained prominence during the seventies when there was energy crisis because of shortage of oil supplies. In India, this led to setting up of Commission for Additional Energy Sources in DST in March 1981. In 1982, a new department was created in the Ministry of Energy by the name of Department of Non-Conventional Energy Resources. In 1992, the same department was converted to the Ministry of Non-Conventional Energy Resources. In the year 2006, it was rechristened Ministry of New and Renewable Energy (MNRE). The ministry has been, since then, responsible for initiating and coordinating R&D, IPR protection, international cooperation promotion and development as well as deployment of technologies for energy sources.

It has been successful in generating tens of gigawatts (GW) of energy so far. Most of it has come so far from wind energy. A large fraction is connected to the grid of conventional sources, but another good fraction has been off grid as well [24]. Substantial indigenous capabilities as well as capacities have been generated in the process. A major amount has also been contributed by the private sector. The policies of the government to support and subsidise these energy sources, R&D efforts as well as increasingly favorable economics have facilitated the process.

R&D strategy of the MNRE has been mainly: development of innovations, improvement of existing indigenous technologies, transfer of technology and assignment of specific tasks to recognised institutes, which is the main operational mechanism. The basket of these energy sources has been quite heterogeneous, in the sense that each category has independent technology subsystems. But energy storage systems are similar. This has been partly the reason for the MNRE to take the help of a network of institutes and bodies like Indian Institutes of Technology (IITs), Indian Institute of Science (IISc), Indian Institute of Chemical Technology (IICT), Bharat Heavy Electrical (BHEL), National Physical Laboratory (NPL), National Thermal Power Corporation (NTPC), National Council of Photovoltaic Research (NCPVR), Solar Energy Centre, Centre of Wind Technology, Sardar Patel Renewable Energy Research Institute and private industries like Suzlon. With increasing deployment, the cost of power has been reducing. It must be remembered that most of this energy push has come as deployment based on components and systems of imported origin. The R&D efforts have not yet come out of the womb of the ministry and most indigenous research in these areas is coordinated by non-scientific leadership in the ministry, though conducted in various R&D institutes in the country.

The Government of India has set highly ambitious goals for renewable energy, particularly solar energy. This could be highly disruptive in many ways. With these types of goals, India will be highly dependent on manufacturing in other countries. Some international players have plans to manufacture the systems within the country, but these will fall short by a huge margin. There are problems related to likely availability of newer technologies of solar panels in the near future which are more efficient, cheaper and faster to make. Even 3D printing has been suggested to make some of these systems [24]. Energy storage R&D is a big associated issue in most renewable energy sources. Developments in this area will also have a huge impact on electric vehicles which can influence the oil import bill as well as cut automobile emissions and related pollution. In view of these, it is imperative that efforts are stepped up on R&D and gradually phase out dependence on imported technology. This requires setting up of a national R&D centre for these areas of science and technology. This will facilitate better coordination of research, efficient transfer and assimilation of foreign technology, act for proper channeling of funds to existing network of institutes and incubation of startups and collaborations with the private sector.

Institutes of STEM HE and R&D

There are several Indian institutions of STEM HE, which have grown in stature globally. IISc and IITs, though predominantly institutes of higher science and technology education, have acquired great reputation both nationally and internationally as educational as well as research institutions. IISc, Bangalore was founded in 1911 and had a series of directors who were of British origin. Sir C. V. Raman was the first Indian director. In the mid-fifties, it acquired the status of a deemed university. It is held in high respect by various international educational institute ranking agencies. It has evolved to be an excellent institution of research. It was the first institute in the country to initiate a doctoral degree in engineering. The Society for Innovation and Development was set up in 1991 as an interaction arm to establish a link with industries. Since then, many initiatives have been taken in the area of IT and electronics in collaboration with several industries like Nokia, Honeywell, Sun Microsystems, among many others [8]. It has forged links with several other industries both Indian and international. The institute also has successfully incubated several entrepreneurs and has generated excellent human resources for higher technological research. IISc is among the top few institutions that have good linkages with manufacturing industries.

Similarly, all IITs have made their distinctive mark in various R&D areas, besides creating excellent human resources for R&D. A major chunk of these resources has benefitted several developed countries. As far as the Indian industry is concerned, some of these institutes have shown potential to generate startups in high technology areas. These could seed enterprises and manufacturing activities and can be a boon for the national mission of 'Made in India.' These institutes have high brand value. They admit only about three percent of applicants vying to get admissions. There are demands from Indian students that one such institute be set up in each state. The Government of India has in principle agreed for the same. A major difficulty that the planners are facing is unavailability of qualified faculty for these institutes. The already existing IIT's are facing a faculty crunch. Bringing a larger fraction of Indian students closer to higher STEM education will give a big boost to human resource development for the manufacturing sector. Several IITs have been roped in for high technology defence projects on one hand and missions like cleaning of river Ganga on the other.

The government has set up five Institutes of Science Education and Research (IISER). Within a decade, they have acquired good rankings in international annals. What is often overlooked is that a substantial fraction of students from these institutions migrates to distinguished universities abroad. Through its inability to retain talent within the Indian innovation system, India is indirectly funding technology and higher education empires of developed countries. These institutes (IITs, ISERs and IISC) could have been the backbone of mission 'Made in India.' Despite these talent losses, these institutions are still India's best bet at the moment.

Indian STEM HE thus has the potential to boost hi-tech manufacturing, but remains poorly harnessed presently for multiple reasons.

Mandates of Scientific Research: Need to Revisit

Scientific research in India is aimed at fulfilling several national goals and most of the government research labs are geared around a specific mandate. Science and technology could impact the nation and its citizens in multifarious ways. These could be strategic security, health of its citizens, food security, energy security, acquiring competency in frontline technologies, developing rational ethos in public life, industrial progress and many more. How the growth of science and technology in India influences each on these parameters?

The mandate of most Indian national labs are implemented by means of several short- term and long-term projects. While basic sciences need not have well defined project-based deliverables, applied sciences and technology development projects are supposed to have more clearly defined goals and deliverables. There is a large variety of projects, some of them are completed within a five-year period (Soviet style five-year plans for economic growth which have been abandoned recently), while others run for several plan periods and form part of a bigger programme. After the end of the Planning Commission in 2014, the scheme of things has changed only cosmetically, as far as R&D is concerned. To implement these projects, most of the human resource base is from within the Indian universities and institutes of higher education. These labs can claim several achievements worth talking. In the national context, all these achievements are s 'first of their kind,' but the question often asked is, whether this is the best it could have been. The fact of the matter is that many of these organisations have become monolithic and have no competitors within the country. It is a matter of debate how they compare with their international counterparts. Is their output satisfactory from the point of view of demand from the stakeholders like user industrial groups or the Indian public? Besides this, they have to be responsive to the demands on them from their future potentialities.

There is a gigantic task which is being expected from some of these organisations. The first question is the time frame in which solutions are expected and the second is, do they have the right kind of human resource base in place. An overload of expectations sometimes is the outcome of a mentality of treating R&D as an off the shelf item. There needs to be an appreciation of the need for sustained and well-planned efforts for building up for the sake of their mandate as well as contributing to building up the national industrial base and other goals of public good. Here is the need to have a structure (R&D-manufacturing) which is balanced and stable to have a broad base pyramid of growth instead of an inverted one.

Besides their existing mandate, how are these research labs geared for fulfilling their "responsibilities" towards mission 'Made in India?' This mission calls for an effective response and contribution from the scientific research and technology development community. This mission is directly linked to economic well-being and quality and quantity of jobs for the citizens. How is government funding various agencies and assessing how they are performing is a difficult exercise for a bureaucrat (or even 'scientist bureaucrat') in a government department. Many a time, these funding exercises are influenced by perceptions, urgency of demands or some kind of networking between scientists and bureaucrats/political leadership. Figure 7 shows distribution of funding between significant scientific departments of the Government of India [5].

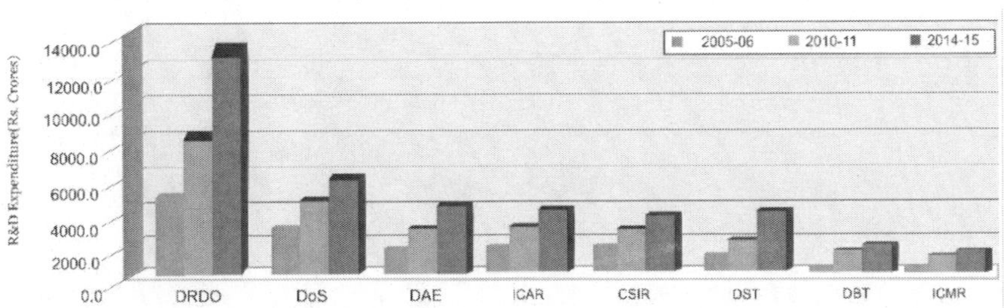

Figure 7. Distribution of annual funding between major government R&D spender organisations from 2005 to 2015 - 1 crore corresponds to ten million (acronyms explained in text).

As can be seen, most of the departments keep getting more funding every year. What about their performance outcomes? The overall record of the country in terms of publications seems fair. Figure 8 shows growth of scientific research publications for the country [5]. The growth rate from

2009 to 2016 has been about 13% as compared to the world average of about 5%. But this tells part of the story. The worth of these publications can be counted based on the number of citations and impact factors which themselves have limitations of their own. This record is much less impressive.

More important for this discussion is the record on patents and their conversion to commercial production. Additionally, a large number of these patents are from MNCs in India that usually use these patents achieved with help of local workforce for manufacturing in their own country. Indian academic output conversion to corporate is quite poor as compared to the global average. If one examines the contributions of the top four countries in specific disciplines and their fractional contribution to the global output in that subject, it can be seen that USA holds a dominant position in most disciplines and many developed countries perform well in most subjects [1]. India is counted in only one discipline of chemical engineering. How different organisations perform in terms of publications and how they compare globally is of interest. From India, contributions of mainly CSIR are noticeable. There is need to keep in mind that many Indian research organisations are involved in strategic research and do not publish liberally. Thus, the complete picture may seem hazy in some ways.

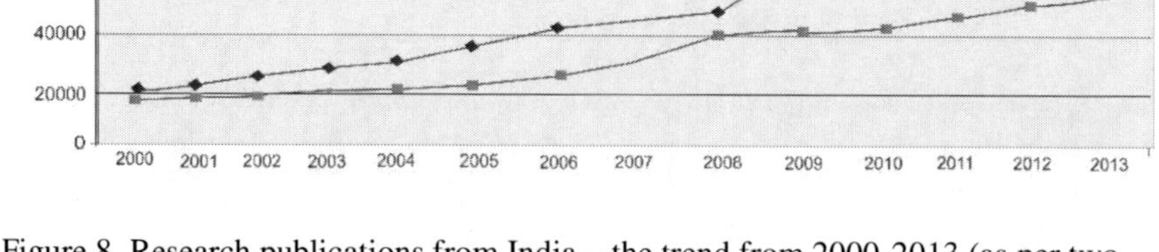

Figure 8. Research publications from India – the trend from 2000-2013 (as per two different source, explained in the text).

Evidently, more important than all these statistical indicators are the requirement from these research labs to be contributing to missions like

'Made in India,' which may acquire strategic worth. Research directions of the Indian government-funded labs and universities usually do not have motives of commercial exploitation. If India wishes to boost mission 'Made in India,' it needs to analyse and be ready for technology transfer revolution of the kind affected by Bayh-Dole Act and Federal Technology Transfer Laws of America in the eighties. There is need to have a fresh look at responsibilities of a large number of its S&T labs. Development of national industrial base and human resource also should be their collective responsibility. It is possible that in some cases their mandate also can be redefined – enlarged in some cases and truncated or modified in the others. Linkages with the end users have to be suitably incorporated and strengthened. Relationships with international institutes need encouragement based on merits of the case and that should be reflected in this proposed reorganisation. This way, strategic sector R&D can be segregated (or restructured) from those in basic sciences so that basic scientists do not suffer from sanctions of any kind. Linkages between various national labs (with overlap of working areas) can be promoted to avoid duplicity of efforts and harnessing synergy. The efforts towards indigenisation need not be indiscriminate but along a carefully charted course – maximising benefits. Some of these guidelines can lead the way for the exercise of mandate redefinition of many of these labs – a disruption with a purpose, with the mission of conceived, designed, developed and 'Made in India' at its centre.

Culture of Research Labs

Scientists, who have worked in different national labs across several countries, relate their experiences to be unique in each lab and of widely different types. Some scientists, with proven competence, do not survive well in an alien environment and leave the otherwise reasonably good organisation. India has seen several cases of immigrant scientific and technology workers who did not do well in India but flourished well when outside. While some of the factors are related to a country as a whole, other experience are attributed to cultural differences of different organisations, besides others. These differences are noticeable even in scientific organisations within a country. University research labs and national labs also have distinctly different cultures. The evolution of culture of a research lab is a long-term process. This bears predominantly the stamp of its founding leaders and also how these labs evolved. These cultural differences are known to have an impact on individual performances as well as that of the organisation as a whole. This can also influence talent attracting and retaining capabilities of an organisation. It may be easy for an organisation to attract funds, particularly from government sources, which is a function of its mandate, but attracting the best quality talent is not always easy. Some organisations do not give importance to this aspect and repent later. How does one define the culture of a lab? What are the various defining features? In the following work, some of the predominant features are examined briefly. How many of the above mentioned scientific research labs pass the culture test that attracts talent, besides good performance, is the moot question.

What is Rewarded?

Only those types of activities and individuals are known to prosper in an organisation who are encouraged and rewarded. So, it is very important for an organisation to make careful selection of people who are rewarded and chosen to lead. It has a ripple effect on rest of the working people. Even a few bad decisions in this context can have a bad influence on the overall morale. Next in the list is whether the organisation encourages basic sciences more than technology development or vice versa and promote or inhibit their linkages? Is the organisation satisfied with only product development of the kind that does not yield patents? Are scientists propelled to be in frontline of international science and technology? Does the organisation encourage collaboration with national and international peers? How much originality is respected? Rewards and incentives can sometime have the opposite of desired impact, if these are distributed in an unjust or opaque manner to individuals, institutes or even disciplines. All these are among the acts of a scientific organisation over a period of time which go a long way in shaping the culture of the laboratory.

How is Failure Looked Upon?

In an R&D organisation, genuine failures are not expected to be discouraged. But there should be a clear distinction between failure of a proposal attempted for the first time in the world and those which have been already achieved elsewhere, implying requirement of gradation of challenges involved. There could be nicely strategised but badly executed projects but also those which are poorly planned and executed. There is strong need to look upon failure with due respect if it is in the domain of normal well studied risk. Lastly, the most important question: Is the person sitting on the seat of technical judgment competent enough to assess?

Rigid Hierarchy

How strongly hierarchical a scientific organisation can be? There has to be a compromise between discipline and encouraging creativity in S&T environment. R&D organisations cannot be run on the lines of administrative units. If it is too rigid, then young creative voices will be stifled and many 'could have been' products (or ideas) are killed in the budding stage by a strong headed immediate superior. If there is accessibility to the top or an alternative superior, then these problems are better taken care of.

Project Sanction, Execution and Review

In India, it is quite often alleged that there is unfair distribution of funding and support to projects and inappropriate performance appraisal of individual scientists as well as projects. There is plenty of "crony sanction" and "crony review" of projects and individual scientists by handpicked experts and peers. Also, while sanctioning projects, availability of appropriate human resources or other inputs is ignored either out of bravado (of ambition) or under compulsion (lack of alternative). These factors do affect the progress of the projects and the morale of working level scientists. These delays and consequent cost overruns do cause a dent in the image and credibility of scientific institutions and the community as a whole, which is a more serious loss.

Emperor is Naked

Quite often, young (or even senior level) scientists lack in courage while speaking to superiors and it is particularly so when the truth is unpalatable. One is reminded of an old folk tale which can of course be retold in modern day contexts which arise often. It is a well-known story, but it needs to be retold here in full.

> This is the story of an emperor who was fond of novel things and appreciated fine arts. He used to encourage his ministers to attract the best talent in those subjects. He was well-known all over the land for his propensities. Learning about this, once a tailor approached him with the request to make a fine dress for him. This will be of the kind nobody has ever stitched or seen. There will be only a few people who can appreciate such fineness. The emperor got excited and started enquiring about how this was possible. What does the tailor need for this task and how much time will he take to stitch such a dress. The tailor replied about the money and other resources that he will require. He told it can be finished in about a month's time. But he put a condition that he will work in an isolated environment and will not like to be disturbed and all his conditions should be rigorously adhered to. There will be a window in his room through which he will interact with the outside world. The emperor agreed to all his conditions and the work started. After a few days, the emperor sent his emissaries to find out what was the progress. They found to their dismay that the tailor was moving his hands in the air as if handling a fabric. They were so overwhelmed with his hype and thought that they were not competent enough to see through the fabric. They returned and mentioned the hectic pace at which work was going on. Few days later, he sent some of his senior ministers. They were also quite shocked but again taken in by the hype and reported similarly to the emperor. Few more days passed and now the emperor himself wanted to inspect. Again, he went

along with his ministers and saw through the window. He was also quite surprised at the frantic activity but not the fabric. He thought since his ministers have already testified to the progress, perhaps the fabric is so fine that he does not have an eye for it. He also came back satisfied with the progress. Few days later he was informed that his dress is ready. The emperor was called and again the tailor made actions appearing like making the emperor wear this new dress. The emperor wore that superfine dress and wanted to go through the streets of his capital displaying the finest new type of dress. The procession started from public square. The public started whispering about the new type of dress. Most of them wanted to speak out about what they felt but either out of respect or fear of the emperor or hiding their lack of an eye for fineness, kept quiet. The procession continued till it came across a group of small children. One of the children immediately shouted, "The emperor is naked" and rest of the group of children laughed loudly. Slowly, the word spread and everybody realised their follies or lack of courage to speak the truth. The emperor finally realised the gullibility of most of the people.

The first thing the story tells us is about the emperor and his gullibility. Admitted that he is not an expert but he needed to have cross checked the track record of the new recruit. Second is about his advisors, who were not up to the task they were assigned and were taken in by the hype and did not rely on their own observation power and did not have the guts to question. They were also sycophants and did not advise the emperor as they should have. The third is about the child who had the courage and innocence to shout aloud an unpalatable truth to the authority. All these have parallels to many modern-day Indian R&D projects, project managers and the sanctioning authorities. The scientist or technologist proposing an impossible project is lacking in integrity, like the emperor's tailor. Then there are very few reviewing experts who have the courage to give comments honestly and fearlessly, like the innocent and courageous child. Lastly, there is the political class that shows its gullibility by agreeing to the tales told by scientific leadership.

Futuristic Technology Development

Does an organisation always work on technology development in a "mission mode" or it follows futuristic leads also? It can dream and plan for future with some vision. Such an R&D organisation is likely to attract talent and may progress better. If it is always working in a "mission mode," then it implies it is suffering from backlog of incomplete tasks and is a sure sign of having bitten more than it can chew. Lack of imagination, appropriate manpower or competence could be some of the reasons for this. These futuristic programmes must follow a timeline where they are able to compete with international peers as otherwise these efforts go wasted.

Linkages between Disciplines

Since most of the national labs are guided by a specific mandate, applied sciences and technology play a key role in shaping them. Each national lab may have a unique mix of their own for basic, applied science and technology development. But quite often this professed linkage between various disciplines is only cosmetic and either basic sciences group moves wayward or applied sciences group is not able to capitalise on the work of their counterpart. Many a time, it is related to poor dialogue and coordination between the two groups or sometimes sheer incompetence. Thus, technology development is not able to make headway. Technology development depends heavily on engineering support to applied sciences groups. Generating the requisite symbiotic environment is quite significant for successful implementation of the mandate related to technology development for manufacturing new products, developing new processes and thus the mission of 'Made in India.' How all linkages between various disciplines operate to achieve project goals is what dictates the success or failure of the organisation.

Performance Appraisal

How individual scientists/technologists and projects are evaluated influences the image of the organisation not only among its own people, but also potential new recruits and peer organisations. This is over and above the influence on performance of the organisation itself. Most of the national labs have a system of evaluation which examines basic technical capabilities, productivity and personal qualities with varying weightages. Employees expect certain degree of fairness and transparency in this process. These factors could be big sources of motivation or discouragement. An often overlooked perspective is that related to working environment. If this is integrated into the appraisal process then it will surely improve the objectivity and ensure better employee satisfaction. What is exactly meant by the word environment? This pertains to what was the type of project? How was it supported in terms of finance and manpower? Whether the project was imposed from top out of necessity or proposed by working people? How it could be graded in terms of difficulty level? There should be soliciting of views of employees about what "systemic changes" could have improved their performance. Besides the reporting authorities, there should be independent and genuinely unbiased experts whose evaluation must be sought in confidence.

In the field of R&D, it is quite often that an experiment or a project simply fails. This is the impression of most of the outsiders and the decision makers or those who approve funds for a project that in R&D, failure is not surprising. This in itself is alright, but how many times it is the genuine failure of a scientific experiment and how many times it is the scientist or the group of scientists who have failed. There is a big difference between the two situations. Decision makers have to be cautious in evaluating these results. Was this experiment destined to fail right from the beginning, either because the people were not competent enough or the choice of direction of attack was not proper, or simply the experiment was going against the laws of nature and it was ill conceived? Also, sometimes the assessment of gestation period is not proper or the time of the idea has not yet come. There are plenty of examples in history of science, where very competent scientists were involved and yet the project failed. This strengthens the impression that in R&D, experiments and projects often fail. The probability of success cannot be predicted. There is need for some bit of education on this count. There is need to differentiate between various types of failures very carefully. Failures cannot be an excuse for lack of accountability. In a large number of situations, most of the time scientists and technologists are not working on the frontiers of science and technology and are mostly trying to repeat what has been already achieved in international science long ago. R&D ventures need to be monitored very carefully in cases where large public funds are involved. If the failure is genuine, in all respects, it is an opportunity to learn lessons and that should not be missed. While evaluating scientists or technologists or their proposals, one must go by their track record, situational inputs and effectiveness in delivering results, rather than sweet charming promises. Being effective in getting a project forward is more important than going through the rigmarole of carrying out scientific process, publishing few scientific papers and finally submitting a failed closure report. The organisation should have mechanisms in place to incorporate these situational factors into performance appraisal of individual scientists as well as projects.

International Collaborations

A science and technology organisation flourishes best with free flow of ideas. In fact, technology diffusion has a lot to do with mobility of its working level people as ideas follow talent. While international labs respect Indian talent migrating from India, they are not always enthusiastic about collaborating with Indian institutions/labs. There is a growing technology denial environment internationally, particularly against some emergent economies. If a national lab is able to progress in critical areas and stand on its feet, then there will be more willing international partners. In addition, there could be denials related to strategic technologies or commercial reasons in the garb of strategic ones. Scientific managers of such national labs have to be alert to potentialities and pitfalls of science diplomacy. Progress in achieving its goals and nature of choice of projects can serve the organisation well in establishing and developing international contacts. Basic sciences, by having achievements in frontiers of science, can also play an important role as an interface in international science diplomacy. Most of the times, frontier basic science projects are associated with high technology and by being part of such projects, a nation gets an opportunity to gain on this front. Joining international efforts for gravitational wave detection by India is a good move, the same is true for DAE collaborating with Fermi Lab, USA and GSI, Germany. But in these cases, care must be taken that Indian scientists and technologists participating in these collaborations are capable enough to derive and assimilate science and technology benefits for themselves. A collaborator with a better worldview can benefit better by manipulating the terms of collaboration. Developing countries have to prepare their manpower and ecosystem for the right kind of receptivity, otherwise these international collaborations tend to be of cosmetic value and at the best promote only a few careers.

Indian science and technology today stands at a crossroads. There is need for comprehensive assessment – what and how to reform as well as the kind of redistribution of support is needed? More significantly, how to make it a more effective contributor to national needs like the mission 'Made in India?' As mentioned earlier, research labs are not manufacturing plants, but can contribute in a big way in building new plants, particularly in high technology areas. There needs to be a bridge between research labs and manufacturers. The developed world has made good use of such linkages which exist as part of long-term legacy of being active participants in major industrial revolution. Most developing countries on the other hand have suffered colonialism and consequently poor attention to science and technology, and they need to assiduously build and evolve these bridges between scientific research labs and industrial manufacturing. For strategic sectors like DRDO, DAE and DOS do have manufacturing counterparts in the form of NPCIL, Nuclear Fuel Complex and defence PSUs like Hindustan Aeronautics Laboratory (HAL) and Antrix Corporation etc. But all these entities are in public sector. Such bridges have to expand in breadth and depth, beyond public sector. Several players in scientific research, technology development and manufacturing have to develop symbiotic structures. The next chapter briefly describes the status of such linkages and various types of players of significance.

Chapter 5

R&D AND INDIAN INDUSTRY: BUILDING BRIDGES

RESEARCH LABS AND INDUSTRY IN INDIA

How do R&D labs perceive industry, how they interact and make use of each other is quite significant from the point of view of national industrial rejuvenation. There has to be a strong symbiotic relationship between the two if industrial growth has to be spurred by innovation. Industrial manufacturing in developing countries is poorly linked with scientific research and technology development. For example, all national labs in India employ scientists and technologists in secure government jobs where research is done for either love (by a few) or mercenary spirit (larger majority) with no extra profit motive than a secure job. Majority of them lack the "killer instinct" required for present day innovation and competition-led growth in technology development. Also, many of these labs are bound by frozen mandates with monopoly in areas of their work. It must be remembered that innovation thrives in an atmosphere of competition and monopoly encourages lethargy. Also, Indian science (dominated by government labs) has a strong legacy where research is ostensibly for the sake of public good. In pre-independence India, science was influenced by the freedom movement and the spirit of self-sacrifice for the nation and society. Sir J. C. Bose is a classic example. In 1895, he demonstrated "wireless communication," first in India and then in London in front of Royal Society. He openly declared it to be not for profit, but for the benefit of society. While at the same time, Marconi in Italy, Tesla in USA and Popov in Russia pursued this idea vigorously for the sake of patents and profit — Marconi could successfully commercialise the technology over a period of time. In another example, it was Dr. Shanti Swarup Bhatnagar, founder of CSIR, who could solve several problems faced by Indian industries around 1940, but did not charge any money for this. Though the times have changed drastically, the linkage between technology development and profit-making is not yet established in Indian national labs, which constitute a major fraction of Indian research establishment. Thus, Indian science could not develop a robust culture of patents and profits and Indian national labs could never frame a clear policy regarding "incentives intimately linked to individual profit" for scientists and technologists. In Indian national labs, scientists and technologists predominantly thrive on research papers and not products, particularly not patentable products (or processes). Thus, they are never able to generate a seamless connect with Indian industries that have also got used to the traditional processes (or technologies) and the "opium" of licensed production based on foreign patents and knowhow, instead of encouraging research of their own. Only a handful of them have their in-house R&D labs.

Indian industry and research labs view each other and their possible cooperation in perspectives that are quite contrasting. Scientists and technologists from big national labs, in general, have a very condescending attitude towards R&D personnel from industries for their being guided solely by "plain profit" motive and their poor level of awareness, lack of exposure to frontiers of science and technology etc. They can, in general, only think of using them as vendors or service personnel. They also avoid them wherever sensitive projects are involved except as low-level contributors. As is evident, it requires major effort to build bridges between industry and research labs that lead to coherent approach for the mission 'Made in India.'

R&D Industry Linkages in West

There is an extreme contrast to how this relationship exists in the developed world. Developed countries have a long tradition of industry contributing immensely to research and technology development and also having intimate links with national labs. In fact, industrial research in many of these countries was born in private enterprises. Also, some of these private labs have pursued basic research with great advantage to themselves. Many of these private labs had sustained interaction with renowned scientists. Many of the American national labs opened their doors to private researchers for utilising their unique facilities and services. As an example, Oak Ridge National Lab and Argonne National Lab have few thousand guest researchers per year from private industries on contractual basis.

American industries have an eye for making market friendly products. America pioneered the industrial research in the early decades of last century. During World War II, soon after the development of microwave oscillator — magnetron for RADAR applications, it was Raytheon Corporation which took help from MIT to productionise it for war-time efforts for the US armed forces. They were successful in just one year, thus making the technology genuinely useful to armed forces. There could be several such examples. There is a strong link between American armed forces and their private industry for R&D and production of new arms. India spends huge amounts to import arms and its defence labs have limited capabilities, but are quite hesitant to share their burden with private industry. Indian armed forces do not share their long-term plan in detail with the private sector for archaic laws on secrecy (only recently there has been some relaxation in this regard and licensing of some defence-related equipment for production to private sector). In contrast, American companies are privy to several of their official defence plans as part of contractual information.

One of the lessons from western lab-industry linkages can be seen from the fact that large public sector labs in USA and Canada have to make available their research results, by law, to public within one year of their release.

R&D for Manufacturing in India: Public vs Private

Indian industry by its very nature is quite focused on short-term profit motive, but it does have the requisite drive to commercialise whatever R&D it performs. The Indian pharma sector is good example of the same. All these features are usually absent in case of those working for government national labs. But what Indian private sector presently lacks, in general, is the strategic vision for R&D and talent. Unavailability of talent is a problem with most of the sectors. Presently, Indian industry and even most government manufacturers and PSUs go in for licensed production. The sectors where industries are actively pursuing R&D are automobiles, pharmaceuticals and IT/IT-enabled services, with the pharma sector being the most active among them [25]. Most of these are growing by leaps and bounds. In many other sectors, either nucleating efforts of the government or joint efforts with private sector may be needed to stimulate R&D. This may depend on the sector, its current level of development and type of investment and human resources required. A noticeable trend is that there is almost negligible contribution from public sector R&D in some strategic hi-tech areas like telecom, materials and pharmaceuticals. The private sector spends more on R&D in these areas.

The Gulf between Indian Researchers and Manufacturers

There are several complementary potentialities between private industries and researchers in the Indian scenario, but there is a gulf in their mindsets. Consumer market is better understood by private players. But at present level of technology development in India, the market for hi-tech products and components in India depends predominantly on government-based consumers and their procurement policies. This is the area which has the maximum impact on linkage between R&D and manufacturing, and growth of correspondingly appropriate relationship. This is also the sector which is questionably considered by policymakers to be not too attractive for increasing jobs. Many a time, Indian research labs use a "lack of industrial base in the country" as an excuse for delay or failure of a project. It is difficult to identify what and who is responsible for this scenario. The industrial base is normally built by slow and steady contributions of scientific research and technology development to manufacturing. Government sector research labs work for upliftment based on their own mandate requirements. Private sector by its very nature invests only when there are clear cut demands. If relevant requirements are well formulated by government labs and publicised in advance along with some assurance to private sector (about orders), then gap areas are likely to be taken care of. If they are given a free hand to choose their collaborating (not trading!) partner from anywhere in the world, then better solutions to some of these problems of "industrial base" may emerge. This may help building of national industrial base over a period of time. Many of these processes of information sharing not happening imply that there is either bureaucracy or lack of 'meeting of minds.'

There is need for both government research labs and industries to reinvent themselves for the sake of mutual and national interests. Presently, both industries (and their associations) and government labs have been making only meek attempts for harnessing their technology development for linking research to production and manufacturing. There is need for systemic changes as well as mindset change on both sides. There could be several expectations from both industries and government labs. Indian industries, particularly the big ones, presently think in terms of R&D as a source only of incremental progress for the sake of value addition, quality control and some means for improved testing and characterising facilities or gaining tax advantages. R&D is not yet part of their long-term vision and strategy. They do not have the requisite human resources. They can prepare a vision document for long-term progress in their own field with help from scientists and technologists in government research labs. Even basic scientists in the overlapping field could be involved in this exercise for pertinent human resource development. This is because Indian industrial researchers usually do not have too much exposure to frontiers in respective sectors. The consultative mechanism between industries and national R&D labs could be a periodic affair, result oriented and on a subsidised/commercial basis, with follow up support from government by policy changes. This could be on individual industry level or that of industry associations. Recently, the government has offered to support up to 80% of development cost for highly required defence equipment. This mode of support could be extended to R&D in more sectors, to be gradually phased out, even when the application is civilian in nature like railways, telecom, aeronautics, mining etc. This support could be sector specific but is particularly true for sectors which tend to burden the import bill. In addition to funding, they need talented R&D personnel. In most sectors, industries at the present stage do not offer attractive career possibilities to young scientific workers. If industries find it difficult to hire trained high-quality R&D personnel, then they should be prepared to train their own staff with the help of national labs, STEM HEI, or be willing to have them on loan basis from these sources or MNC R&D centers (or even international setups). Presently, their R&D expenditure ranges around 1% of their turnover — that too for a select few. [25]. If they wish to compete internationally, then this average also needs to shift to international standards. One of the mechanisms to build bridges is for industries to identify good projects (proposals) for joint work with government labs, which could benefit all the stakeholders, viz. industries themselves, research labs, individual scientists and society (or consumers) at large. Some of it could involve multiple partners.

Several steps are required for industries and government researchers to appreciate each other and build bridges. Presently, industries find the bureaucracy of interacting with government labs and Indian university departments daunting. Government labs need to smoothen the process to offer facilities of human resource development of scientific and technical staff of some of these industries on nominal payment basis. More significantly, government research labs need to build their credibility in terms of offering the right kind of R&D knowledge base of potential use to manufacturers. Government labs should be prepared to offer their specialised design, testing, analytical and characterisation facilities again on liberal commercial terms. There could be several such facilities existing in government labs which are underutilised. Such sharing of facilities can be advantageously done when these are costly to purchase or maintain. National labs could help industry to evolve standards and quality control techniques and protocols consistent with international practices. Besides using industries as vendors, fabricators or service providers, they should be prepared to collaborate with industries on suitably selected R&D projects of mutual interest or long-term national interest. Initially they can be the senior partner. The big brother attitude has to be slowly shed off and industries could be allowed to be the leading partner in choice of projects or mode of execution. Scientists working on industry-related problems together with industry need to be incentivised. Those out of this loop (like basic sciences staff) need to be involved in human resource development for industries working in high technology sectors. For this purpose, each R&D institution has to evolve a balanced perspective and policy on development of applied sciences and technology development on one hand and basic sciences and human resource development on the other. Government R&D labs should encourage incubation of ideas with the help of industries and should facilitate some of their own staff opting to make a shift to say a startup industry. This should be treated as a useful transformation from one kind of national asset to another. After initial development of a product or system prototype in a government lab, suitably chosen private industrial partner should be encouraged to take it up. The choice of industrial partner should be made at an early stage of technology development. This will help promote the linkage. Transfer of such tasks should go beyond government manufacturers (or PSUs). Government procurement suffers from excessive regard for lowest quotation and fear of vigilance agencies. This has to be overcome by policy means so that indigenous R&D-based products manufactured by industries can be supported vis-a-vis foreign products. There needs to be a more level playing field for Indian innovators.

This is such a big wish list that it may not sound feasible. High level intervention and political will may be called for to enlarge/amend the scope of mandate of national labs and not look at private industry with eyes of mistrust. They need to be more autonomous to deal with private industries. Proper safeguards and guidelines (without much bureaucracy) have to be incorporated to avoid misuse of this freedom. Many of these tasks related to industrial R&D are supposed to be in the domain of CSIR. Since technology scenario is expanding and diversifying and it may be difficult for a single agency to cope with it on all fronts. It will be desirable that many more national labs and universities be co-opted for this purpose.

BRIDGE FOR 'MADE IN INDIA'

The immediate objective of the government with the mission 'Made in India,' is to make use of Indian workforce and mitigate unemployment. Presently, there are barriers to attain the goals of designing, developing as well as manufacturing in India (or 'Made in India') for various reasons such as inadequacy of existing talent and infrastructure, and difficulties in overcoming the challenge of cutting through existing international supply chains and poor industrial base. This is while MNCs are making use of Indian workforce for some of these tasks. India needs to cut through these barriers and build a robust bridge between industries, researchers and innovators. How some of the fast-developing countries have achieved success in this respect could offer lessons. A frequently quoted example is that of China, where it is pointed out that the country has benefitted immensely from the FDI route in terms of technology transfers. It is not clear how much of high technology progress in China has simply happened on the basis of FDI and how much is indigenous effort. It is also often alleged that much of this is related to technical espionage of western industrial empire. It must not be forgotten that MNCs are careful not to create their own competitors. In this context, it will be worthwhile to refer to a testimony in the American Senate on US-China Economic and Security Review Commission [26]. The summary of this testimony points out that China is gaining just a little as far as technology transfer benefits are concerned. As an example, an Apple phone costing $299 produced in China has the Chinese value addition component of only $4. Its conclusions based on data till March 2011 points out that "most recent data show that US-headquartered MNCs have 70 percent of their operations, make 89 percent of their purchases, spend 87 percent of their R&D dollars, and locate more than half of their workforce within the US economy. This predominant focus on the home economy has persisted over time and changes only very, very slowly at the margin." Such examples reassert the need for indigenous technology-based manufacturing, while steps like FDI and offsets etc. can only be supplementary in nature.

Role of Scientists, Technologists and Managers

Most of the Indian industry is supported in a big way by engineer-managers. Higher the technology of a product, more it is likely to be based on foreign knowhow, licenses or patents. The breed of engineer-managers is growing by leaps and bounds and India is aping the trends in some of the western nations. But it must be remembered that India as a country is losing the number of potential designers and builders at a fast rate. The proportion as well as the role of scientists, technologists, engineers and engineer-managers needs a thorough review.

Indian business schools that have good global reputation can contribute to the process of transforming the scene by generating a community of engineering and marketing professionals for the purpose of replicating the process for indigenously developed products. The process of designed, developed and made in India products poses a challenge to B-School professionals and teachers to inculcate the spirit of taking off from the stage of 'idea to product to market' instead of 'license to market.' Some of the startup incubators associated with Indian Institutes of Management (IIMs) is a good beginning in this direction. So far, predominantly, the initiative has come from management communities to inculcate spirit of enterprise in Indian technologists, but more significant will be the researcher and developer community taking initiatives in this direction. This must be realised that for all their competence in R&D, Indian researchers are not good at bringing an idea to the level of production and profits, exceptions notwithstanding. This community of business managers (or entrepreneurs) may also provide the much needed interface between the Indian industry and research community.

These facts are applicable in general to both public and private sector researchers in India. In principle, private sector is motivated by profits, but still, taking R&D from the stage of idea (or a raw product) to a marketable product is not as straightforward for them too. Even more difficult is conceiving and developing a product first for an unknown and unsure market as this involves bigger risk taking. Soft sectors requiring fewer capital investments can afford such risks and hope to flourish. As can be noticed in the case of IT sector, successful innovation and R&D by bigger players has resulted in a good number of smaller startups. A similar revolution for hardware-based technologies for the sake of the mission 'Made in India' may require a very different ecosystem. Government intervention/investment may be necessary more often, particularly at early stages.

The erstwhile Indian Planning Commission's paper on strategy for manufacturing too had reminded that R&D, technology upgradation and value addition is needed in most of the sectors [7]. Policymakers are aware of such problems, but there is lack of action and accountability. This comment is common to both traditional and modern sectors like steel, textile, food processing, ship building, chemicals, fertilisers, machine tools and many others. This document repeatedly mentions the need for industry-academia interaction [7]. There is talk about barriers to enterprise creation but rarely about barriers to collaborations, cooperative interactions and joint projects between researchers and manufacturers. There are various kinds of hurdles, be it bureaucratic, cultural mindsets or egos of individuals and institutions. Universities, research labs and industries have different goals and accordingly develop different levels and kinds of knowledge and skills. As per the World Bank report, a huge knowledge base differential exists between various research and manufacturing entities in India [8]. Because of poor interaction mechanisms between various research institutions and manufacturing entities, this knowledge base does not get percolated to where it could have been useful. There are distinctive problems related to each sector or pair of potential collaborating partners. For the sake of this discussion, most knowledge bases can be divided into three categories: "new to the world," "new to India," and "new to the entity." The first one refers to proprietary knowledge and patents, the second one is what has not yet been developed within the country but can be acquired from abroad on payment basis or route of mergers between private entities, and the third one is what exists in some of the national labs and higher education institutes (HEIs), but the related private industry or manufacturing PSU is not aware of this. About the third kind, there is ignorance about possibilities and this is irrespective of their will to pay for this. By overcoming suitably this knowledge differential, it should be possible to reap huge productivity benefits. Domestic technology flow itself could be a big booster for the role of technology growth among industries (wider deployment rather than development alone). This also promotes bridge formation between the recipients and donors of technology — a long term gain. There is wide spectrum of sectors and level of technologies operative in different research and manufacturing units in the country and where good overlap exists between many pairs (researcher/manufacturer) or groups. What are the various possible indigenous sources as well as potential recipients of these technologies and knowledge base? What are their roles and constraints they face in this process? Some of these are described below with their potential and present limitations. Each of them needs to develop multidimensional bridge networks between other players. Some possibilities are cited as examples.

Networking All Players

There are several research and manufacturing entities that need to network among themselves. There could be several pairing or multiple player joint enterprises. Some of these players and their linkages are described below. India has a large number of public sector manufacturing industries built in the Soviet style setup since the fifties. These PSUs are the first category of manufacturers run by government institutions operating for variety of industrial sectors. Their goals had been developing indigenous capabilities and capacities in important sectors of the economy. They are involved in wide range of manufacturing. Each of these PSUs face different types of challenges and problems. Collectively, many of them provide a unique national asset and infrastructure. Many of them had done good nucleation of manufacturing activities in their respective spheres. Many a time, these public enterprises had been anchors for taking a prototype from a national R&D lab and productionising that product or even have done a good job of building systems or products with technology imported from abroad. DPSUs have developed some major systems of great strategic value but marred by huge delays and cost overruns. This process might have given them skills to build, assemble, test and maintain, but there was only limited emphasis on nurturing research, design and development capabilities of an improved version. Indian Railways is one such example where at one time it started exporting locomotives manufactured with the help of foreign collaborators. This was the case of 6000 HP locomotive built in Chitranjan Locomotive Works (CLW), when Chinese were at par with India. Within a decade or so, India once again was seeking technology for better and higher power diesel locomotives and Chinese offered to sell the same in competition to Alstom and General Electric. These will now be manufactured in the state of Bihar by the latter two. Lack of stress on localisation and developing capability to upgrade has been a weak point of most PSUs. Besides these locomotives, Metro rail manufacturing is the same story. India is only manufacturing Metro rail, while most of the design work and innovation takes place in design shops abroad. This is when a large number of Indian cities is vying with each other to get the Metro started. The story is likely to be repeated for bullet train and other high-speed trains. This is directly linked to lack of culture for innovation and R&D. The government's recent decision to set up a Railway Research Institute is welcome awakening, but it is too little, too late. Hopefully, this may change the trend. The situation in the telecom sector is no better, where Indian telephone industries, which were established in 1948, and Center for Development of Telematics (C-DOT) at a later date, could achieve very limited success. Once again, Chinese companies are in the list of potential suppliers. It is yet to be seen how much of hardware required for Digital India and BHARATNET programme will be designed and manufactured in the country. There could be many such examples of similar nature.

A large number of PSUs suffers from under performance. Many of them operate in a monopolistic regime. Hindustan Machine Tools (HMT), the watch making PSU which was closed sometime back, is an example. At one time, it used to be the only watch manufacturer in the country. Only smuggled foreign brands competed with it. HMT at one point was not able to cope with the demand of their products. They had built good technological infrastructure, manpower and brand value. This was frittered away for reasons of poor R&D, inability to adjust to new technology (mechanical to quartz), cost inefficiency and inept bureaucratic management.

Very often, the solution suggested to their poor performance is to close down these units (like HMT) or pass it on to some private industry. These are trivial solutions usually offered by an accountant or book keeper, to a very serious national problem. Under pressure, they keep juggling with book keeping to show marginal profits, instead of banking on R&D. For example, some of the pharma PSUs that have contributed significantly to the growth of Indian private sector (pharma industry) in earlier times and are no longer in a healthy state , are suggested to be considered fit cases for closure. It is not based on analysis of the root cause. It is often that their R&D is no longer relevant to present day reality. The scenario suggests that some of them must be revived and encouraged to collaborate with the more experienced private sector to address topical problems. Exchange of ideas, knowledge base on problems can be more helpful.

For PSUs, building up long term national technological competency should get as much weightage as immediate profit, though well-defined metrics for this purpose has to be evolved. Thus, more active engagement with R&D — picking up relevant problem areas of new age context or neglected problems of public interest, initiating collaboration with relevant national labs or institutes of higher education as well as passing on some of their problems to joint consortia with some well-chosen private industries will go a long way in alleviating many maladies affecting many of these PSUs. Many PSUs and DPSUs have common maladies like poor weightage to R&D and technology development, led mostly by bureaucrats expectant of associated perks, short tenures of leadership, work culture of a government organisation and being not in sync with R&D culture and limited autonomy to leadership to collaborate with Indian or foreign research or technology partners in the field. These shortcomings prevail in varying proportion in different entities. Some of them have been trying to develop new technologies but with limited success in developing more critical components. Indigenisation is cosmetic and mostly confined to components and systems of poorer strategic (or commercial) value.

Many of the heavy industries from several PSUs had a monopoly for a long time and in some sectors, they still have. Making them face competition from private or even foreign players on equal terms is the best way to push and persuade them to the path of making innovation and R&D as their cornerstone. Even if they import and license some technologies, that should be with an aim to assimilate these and follow with designing, building or reverse engineering the next scaled up or advanced version of the same product themselves, maybe in collaboration with some other national lab/PSU/HEI. Korea and Taiwan have made very successful use of such strategies. Private sector or sometime even MNCs (in carefully selected cases) should be allowed to enter that arena. Some of these PSUs could be prepared to assimilate frontline technologies accruing from the offsets earnt against major purchases. But this will be possible only with the help of an appropriately chosen R&D partner or some national lab or HEI.

Lastly, these PSUs have a role to involve MSMEs for many of their tasks with an aim to build up national skill base. Developing human resources beyond their own walls should also be considered an important goalpost. Any quality control issues of products from these MSMEs should be sorted out with suitable training to manpower from these units. This way they can complement each other. Thus, PSUs will be able to focus on their mainline job.

Public Sector Research Labs

Indian national labs are repositories of knowledge covering several frontiers of science and technology, besides achievements related to their mandate. In this context, it is worth recalling the erstwhile National Planning Commission's strategy paper on manufacturing suggested in the context of industry-academia interaction: "align the goals and annual planning processes of central research institutes with that of industries through industry associations" [7]. It is noteworthy that except for a few cases, such interactions and dialogues so far had been mostly bureaucratic and not very effective.

Growth of Indian pharmaceutical industry is a very good example of fruitful interaction and flow of knowledge base between central research labs and private sector industries. This model rested on nucleation of process at central labs to be followed by scaling up at private sector industries and remedial action in case of any developmental difficulties back at government labs. The Indian government has set up several pharmaceutical industries in public sector. Two such undertakings — Hindustan Antibiotics and Indian Drugs & Pharmaceuticals were responsible for the first development of bulk drugs in the country. Also, many of the national R&D labs like National Chemical Laboratory (NCL), Indian Institute of Chemical Technology (IICT) and Central Drug Research Institute (CDRI) had done good foundational work on the process chemistry of many drugs and this technology was passed on to private sector labs that could scale it up. Also, many of the senior researchers from these labs (including Dr. Anji Reddy of famed Dr. Reddy's Labs) moved over to private sector labs along with their knowhow and expertise. These government labs also helped solve many teething problems faced by these industries. Another facet is that much of basic level human resources required by these industries were mostly trained in government pharma education institutes. The impressive growth of private sector pharma companies owe a lot to Indian national labs.

Indian agricultural research is mostly performed in ICAR and state level agricultural universities. This resulted in ushering of Green Revolution and India was in a position to ward off food scarcity. A huge knowledge base has been passed on to Indian farmers. But besides this, many of the hybrid varieties developed by private R&D has its foundation in strong public sector R&D. For example, crops like pearl millet and sorghum [27].

DRDO is another R&D lab conglomerate that has been transferring R&D for the sake of manufacturing. One of the recent and reasonably successful examples is that of light combat aircraft (LCA) which was developed by Aeronautical Development Agency (ADA, a wing of DRDO), and was accepted after a lot of tests by the Indian Air Force (IAF). This development story has lessons for R&D labs and manufacturing entities (HAL in this case). Another first is transfer of technology of pilotless aircraft LAKSHYA to a private player that was involved with the project right from the beginning. This offers advantage of further development and proper deployment applications and marketing. Each of these sectors offers interesting and useful lessons for Indian research-manufacturing ecosystem to advance in future.

Presently, several public sector labs already have some mechanisms for transfer of technology. This is mostly in the cases where a new product or technology is fully developed and is then offered as a package to prospective users. Mostly these are small ticket products, which could be of utility to public less by design than by coincidence of dual use, employing spinoffs from technology developed as per the mandate of these labs. This may already be happening to some extent. Presently, this process of technology transfer from labs to industries is generally ad hoc and poorly organised. Proper planning and involvement of recipient of technology at an early stage in the process of technology development will enhance the stakes for both partners and benefit the process immensely. DRDO and Federation of Indian Chambers of Commerce (FICCI) have launched the accelerated technology assessment and commercialisation programme which aims at transferring some of the technologies like insect repellant, water purifier, explosive detection, ready to eat foods etc. For this purpose, FICCI has established a Centre for Technology Commercialisation which takes up all activities related to technology assessment and possibilities of commercialisation.

The nuclear submarine developed indigenously is a fine example of collaboration between multiple public sector research labs and several private industries. Only some of the strategic applications like this bring several entities together. Such collaboration model could have been more widespread for a variety of non-strategic but challenging applications.

For efficient and more intimate interaction, a significant issue is calibrating modes of knowledge base transfer. There could be various modes in which such technologies or related knowledge base can be transferred. These could be by way of advisory or consultancy nature, giving the benefit of expertise or experience. This is pursued via the route of memorandum of understanding (MOUs). It could also be training manpower of the recipient in some experimental techniques or lab or shop floor practice, S&T services, like test, calibration, measurement, qualification/certification, modelling code or numerical analysis/simulation etc. These could be required by several industries. Specific technologies developed by R&D institutions or even know-how could be transferred. There could be contract R&D at the request of industry on a specific problem. Lastly, there could be joint venture with the industry or a group of industries, on problems of interest to either. These are most probable modes of sharing and exploiting knowhow. For any of these modes, implementation and equal commitment of both partners till the last stage of commercialisation is the key. For the sake of maximum impact in these knowledge sharing ventures, the goal in developing countries should be to minimise the layers of bureaucracy involved between the individual (or group of individuals) responsible for transferring the technology and the user group in the industry. Issues related to joint ownership of proprietary rights are relevant in some of the modes mentioned above and if any licensing of intellectual property is involved, it should be made as smooth as possible. This may require a radical departure from existing mindsets and practices. Such policy should be enumerated at the national level and not at the level of institutions for the sake of accelerating the process. Even norms of human resource involvement need to be better defined for smoothening the process. These could be the major steps required to allay any doubts or fears of any of the collaborating partners.

While talking of knowledge base, it is not purely technologies for manufacturing which need to be considered but also imparting education and skillsets with potential to catalyse assimilation of latest technologies and related manufacturing, which need to be considered. The task of building up this base is so huge that it requires all avenues to be explored to alleviate the situation of 'technology ignorance' of all kinds in manufacturing entities, particularly MSMEs. This may have proportionate productivity dividends too. Many of the Indian national labs are eminently suited for taking up this task, but may require their infrastructure to be tweaked suitably. This is not part of their existing mandate and may also require suitable amendments by intervention. Also, more important for them will be to interact with MSMEs rather than with big industries alone through industry associations. These MSMEs need to be sensitised to possibilities of technology upgradation of their processes or systems by institutes within the country. There are several sectors which are otherwise significant but poorly supported by corresponding research institutions. A suitable mechanism of interaction between the lab and MSME, like scouting for example, has to be worked out. This way several technology upgradation programmes of these MSMEs can be made more meaningful.

Another area of activity which can stem from national labs and could be helpful for 'Made in India' is that many of the national labs have some unharnessed technologies, products and techniques and were never considered in the category of spinoffs. These could be exploited for domestic consumption or even exports in some cases. Many of these have not yet been harnessed because of reasons like poor branding, lack of a proper business model or economics not worked out, scaling up difficulties as private industry was not coming forward, fearing low demand of a new product (or poor after sales support from supplying lab mired in bureaucratic machinery) and lastly lack of political will in case of strategic products exports. Many a time, it could be plain 'bureaucratic inertia' or 'hesitation of last mile effort' of a scientist/technologist. There can be vast opportunities and they require to be explored with due diligence. A more significant mode for such technologies to be harnessed in the present context may be to encourage startups with the help of involved researchers themselves. Barriers to such enterprises can be overcome with greater emphasis by establishments and stronger will of researchers willing to spinoff a startup. There is need for the government to ease the process of spinning off startups from national labs rather than products.[9]

The knowledge base from these labs, which is neither strategic nor proprietary, needs to be converted to an open source one for better dissemination. These should be publicised widely on websites of these institutes and audio-visual media for possible productionisation/scaling up opportunities by industries/MSMEs. This will stimulate competition between them of the right kind. Such publicity should be mandatory soon after claims about development of any technology in these labs. A suitable

---

[9] There could be several examples of unharnessed or poorly harnessed indigenous technologies worth mentioning. For example, an electric vehicle based on say lithium ion battery is yet to be manufactured and deployed on a large scale, while technology for such a battery has been developed years ago by ISRO costing a few hundred thousand INR. Biogas plants are another example – this has been developed by Bhabha Atomic Research Centre (BARC) and some startups have taken it up but has not been multiplied in numbers. Lasers, developed indigenously by several institutes, meant for applications in micro-machining, therapy or environmental applications, could not be harnessed for reasons of last mile shortcomings. Water purification is required in a very large variety of situations in India – both rural and urban. Technologies for these requirements based on membranes or radioactivity-based techniques exist within the country. There could be a large number of such examples – both for industrial or other applications. In most of such applications, the costs could come down by dint of large number of users. On high technology side there are super computers which could be built in larger numbers for many research and educational institutions. There are products like Brahmos missiles, satellite tracking stations, imaging centers and even nuclear reactors which could be exported to many countries. There are several other examples like that of radio-therapy units, analytical instrumentation for large variety of applications. There could be several reasons like quality of products, poor assurances on stability of supply/services or even lack of political will.

model has to be evolved to share responsibilities and profits, if any. They will be encouraged by proper backup support.

Private Sector Industries and R&D

Most of the Indian private industries have developed an entrenched belief that licensed production with foreign partner supplying the R&D component is the best and most profitable route. Some economists in developing countries have reinforced this impression, based on considerations of short-term gains. Such behaviour has led to a conservative mindset ignorant of futuristic possibilities. There are several other reasons for this state of affairs. Besides normal hassles of starting a venture into high technology area, they have to pick and choose the right kind of scientific or/and technology manpower and this has to be done in a seemingly talent deficit market. It is in the case of manufacturing, for example, that R&D into new age (or hi-tech) manufacturing is yet to become a norm in India, though some of the big industries have shown inclination to invest big by recruiting high cost foreign talent for the sake of frontier R&D.

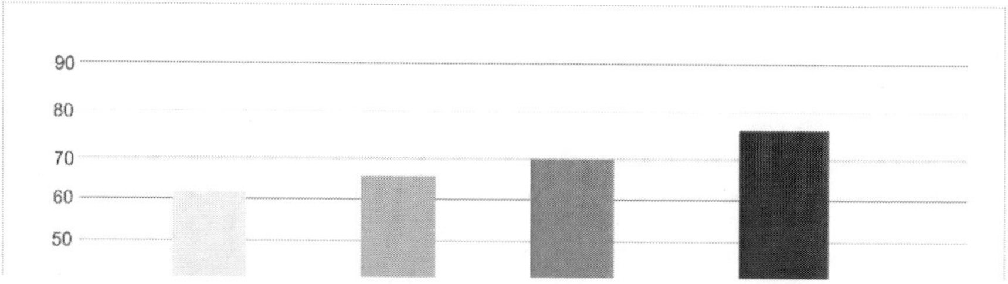

Figure 9. Increase in total R&D expenditure of the Indian pharmaceutical industry (in billions of dollars) from 2014-2017.

It is interesting to notice the progress in expenditure on R&D by private sector companies. Among the top spenders on R&D are companies from IT and pharma sector that have tasted the fruits of R&D already. More than two third of Indian companies listed on Bombay Stock Exchange do not report any expenditure on research and development. Tata Motors is among the top 50 corporate R&D spenders in the world, but most of this is done in its labs in the United Kingdom. These examples indicate the attitude of Indian corporates towards R&D. Figure 9 shows increase in expenditure on R&D by Indian pharma companies in the recent years [28].

Some of them are spending up to 9% of their sales turnover on R&D which is comparable with the global average. This has come about mainly as they started perceiving threat from MNCs after the new regime of patents. The response of private industries in the pharma sector to these challenges has been better than many other sectors. Indian industrialists need to pick cues from MNC R&D centres being set up in India. These centres are pumping billions of dollars per year indicating their banking on the Indian research worker who is costing them one fourth (on an average or even less) of an equivalent American researcher [8]. They intend to make use of this research for manufacturing back home. It is evident that indigenous manufacturers are under exploiting the opportunity to harness Indian researchers. This is indicative both of poor vision and lack of technology leaders among industrialists. They have learnt to prefer alternate routes of profit. They are usually discouraged to invest in R&D for various reasons. Nucleation of R&D in any private industry is the most difficult part, for reasons pertaining to funding, talent and R&D infrastructure. Equally challenging is the last mile of technology development and marketing of a high technology product. For private industries, the next best option to investment is to find a good collaborator. A good possibility of collaboration is Indian national labs and institutes of higher education or universities, provided these generate knowledge base of the right kind. How to make progress in this respect? TheWorld Bank report [8] points out that receptivity of new technologies by industries depends on factors like size of organisation, education level of its employees and management, exposure to type of customers or international interactions etc. Exposure to better education and in-house training of workforce is a possibility and interactions of private manufacturers with national labs pertinent to that sector will surely enhance such possibilities. Collaborative efforts between the defence, aerospace and nuclear labs with private industries like Godrez & Boyce, Larsen & Toubro, Tata Power

SED and many others have benefitted Indian national labs too. A few recent examples include the Indian Ministry of Defense giving major developmental projects like battlefield management systems, tactical communication systems etc. to private sector consortia. This is being expanded in a major way through a process of selection of strategic partners. It will be worthwhile for the government to envision and share long-term perspective plans in several more areas with industries as research and development readiness on part of private sector cannot come about overnight. Stability of policies is one of the major attractors for investors and innovators. There could be shared objectives but well-defined distinct roles as either partner needs clear definition of the same for the sake of mutual benefit. The policy framework should thus be made flexible to the extent of providing equal opportunity to various players, big or small, young or old, and capital rich or poor. The present (proposed) system of selection of strategic partner for big ticket investment in defence sector addresses only one aspect of the problem. This policy does not bother about innovations happening in small, young and relatively poor startups or MSMEs. This will be required in future as more and more startups backed with innovative solutions to old problems emerge. Any collaboration or incentivisation model has to respect innovation more than capital.

Many progressive industries are keen that government creates a framework where large industrial consortia can work jointly on technology of their interest with academic institutions. This was one of the pre-budget demands by some industry groups from the government, in preference to the usual demand of tax incentives. Implementing such a move shall definitely encourage government labs to work on problems of greater relevance to manufacturing. Occasionally, this may also lead to deviation from their original mandate. Presently, private industries enjoy 200% weighted deduction on their contributions to universities and academic institutions. More important than donations is to have shared goals on technology development. But the legal framework, with liberal mindset, for facilitating such collaborative ventures may have to be put in place.

This involves minimising bureaucratic hurdles to two-way movement of human resources. A large number of government labs are hindered by security restrictions, thus movement to and from national labs with higher security ratings may need to be suitably vetted by the security agencies. This process needs to be smoothened as much as possible.

University Labs and Industry

Many of the Indian universities, because of their potential to supply appropriate human resources, can contribute in a major way to manufacturing industries. Their main job is to educate and the next job is R&D. As it is, only about 15% of Indian universities and institutes of higher education are estimated to be involved in R&D. Only 4% of total national expenditure on R&D reaches them [17]. It is too meagre for the extent of enrolment in them and there is a huge scope for improvement. While the government, both at the Centre and in states, can do their bit in improving this situation, science and technology departments of universities should also come forward by reorienting their R&D directions and taking responsibility of augmenting their resources on their own. They have to reinvent their role for the new goals and situation. Keeping the goal of the mission 'Made in India' in mind, the first and foremost demand from universities is to inculcate the spirit of innovation and enterprise, to go beyond basic rote learning and training students for real life situations. Private industries involved in latest technologies may seek new knowledge base and research results which universities can facilitate, provided they choose to work in R&D areas of interest to industries. This is not the case presently for most students, researchers and universities.

Most of the research grants to universities are provided by government organisations and contributions from private industries are rare. Also, many more industries will need to venture out to hi-tech or more non-traditional areas where university knowledge base can be more useful. The scenario in India has to be seen in contrast to the situation in America in the eighties following Bayh Dole Act, when university researchers came forward in large numbers to transfer proprietary technologies to industries. This is indicative of their right choice of directions of research even prior to this legislation. Indian universities and HEI on the other hand feel uncomfortable, for various reasons, in working on problems of relevance to industries.

It is Dark in There

> Mullah Nasruddin lost his ring in the basement of his house. He looked for it but could not find it. Then he went to the front of his house and started looking for it there. His neighbour asked him,
>
> "What are you looking for?"
> "I lost my ring," he replied.
> "Where did you lose it," enquired the neighbor.
> "I dropped it in the basement," Mullah replied.
> The neighbour was quite surprised.
> "Then why are you not looking there,"he remarked.
> "It is quite dark in there and I cannot see a thing. That is why I am looking for it here,"the Mullah said.

In R&D, both basic sciences and in technology development, it is quite often that people prefer to work on problems which are relatively easy, ignoring those which could be imaginative, original or essential. Another habit is to work on trending problems and these are set by scientists and technologists in developed countries. At least many of the R&D problem areas in applied sciences and technology can be identified from the socio-economic situation of a country. For example, in India, tropical diseases research should be of greater interest as compared to say ozone hole and its relation to skin cancer. Renewable energy resources like solar energy should be of much greater relevance to scientists and technologists of tropical countries near equator as compared to those of cold countries near the North Pole. Then some R&D problems are able to attract funds, fellowships, scholarships or prizes more than others. Researchers prefer working on a fashionable problem with less effort and where only incremental results are expected.

It should also be the responsibility of scientific advisors to point out the right research directions to political leadership or funding agencies. Research workers must also bring light to where darkness is, rather than keep searching where there is already light.

There is need to provide a legal framework where university researchers, both students and professors, can exit the system, work on technology-related enterprise for some time and come back, in case of failure. It is hoped that with new thrust on missions like 'Made in India,' improved ease of doing business and encouraging innovation by the way of Startup India, things may improve for better and more potential techno-entrepreneurs may shed inertia. There could be more marriages between science and enterprise. A synthesised pool of knowledge and experience can be much more effective than when they work independently. This is as true for individuals as for organisations and companies.

Governments (both central and state) need to help universities in setting up business incubators in university campuses. Presently, some of the IITs are main sources of startups (with some exceptions). This should extend to majority of Indian universities, only then India will be able to capitalise on richness and diversity of its vast manpower. This could have a multiplier effect if it develops potential to attract R&D talent from some of the other developing countries, like its hospitals are attracting patients from there. Presently, the potential to attract foreign students is very limited and this impacts international rankings as well.

Governments are facilitating setting up of national manufacturing zones to accelerate manufacturing activity. These should be redesignated as national innovation (besides investment) and manufacturing zones. In these zones, universities should be allowed to set up some suitably chosen science and technology departments or institutes. Regulatory norms should be relaxed for such off-site campuses. The government has recently relaxed these requirements for tier I institutions, which is a good beginning. Choice of these departments will depend upon the type of manufacturing industries in their neighborhood. Mutual interactions between them could prove useful for setting up new enterprises and improving productivity of the existing ones. In this context, it is worthwhile to recall how Stanford University in California started offering extended lease of university land to hi-tech companies in the seventies and this fructified years later and Silicon Valley was born. Mutual interactions between researchers and hi-tech industries could have exponential gain factors.

On the other side of the spectrum, universities should take help from manufacturing industries too to design curricula as well as evaluation methodologies for real life situations. Students should be encouraged to do projects in these industries and could be absorbed later on in the same. Private universities can be more progressive in this context, if they are freed from the large number of regulatory constraints. Many of the universities do have sufficient funding to initiate R&D and attract talent. Many of them have started several projects on disruptive and futuristic technologies.

MNC R&D Centres

A large number of MNCs have found it attractive to set up R&D centres in India even when they are not equally attracted towards corresponding manufacturing activity in the country. As can be seen, MNC R&D centres are relatively simpler ventures and can be in standoff nature, while manufacturing has to deal with several other related activities like sales, marketing, linkage to supply chains, infrastructure and logistics issues, competition from other players etc. These research centres in India belong to various countries like America, France, Germany, Netherland, Korea, China etc. More than 70% of these centres are from America. They have invested more than 10 billion dollars so far and are growing rapidly. Presently, there are reported to be a few thousand such centres. They employ a few hundred thousand local employees and are expected to reach about a million by 2020. Many of them have up to 70% of their staff as researchers. They recruit from tier I, tier II, and even tier III universities, as well as HEIs. They offer in-house training to young recruits and find it still profitable. But for their collaborative research, they mainly interact with tier I institute and structure their programmes around skillsets of local students. Their mode of operation is either to have a subsidiary of their parent R&D centre or to buy a domestic startup. Many of them have converted their captive BPO centres for engineering R&D. They are working on many disruptive technologies like IOT, 3D printing, wearable devices etc. There are others that are exporting engineering services, besides software-telecom, automotive, semiconductors etc. But for most cases, major manufacturing is done in their parent country.

It remains to be seen how these R&D centres will impact Indian domestic R&D and manufacturing activities. The first likely impact will be depletion of available talent for domestic R&D institutions. But there will be advantages related to exposure of Indian researchers to international MNC research ecosystem, besides of course employment. Over longer term, this may also stem the tide of immigration of STEM talent from India. Finally, it is up to the domestic R&D and manufacturing entities to learn from these experiences and leverage partnership with them, as mentioned above. There is need to carefully understand their mode of operation and how they leverage the potentialities in different countries, adapting themselves to diverse situations across the globe and retaining the objective of benefitting their home country the maximum. This requires Indian institutions to understand the motives of these MNC R&D centers. There could be variation in different centers based on their country of origin and area of expertise. It can be seen that there is a wide diversity of reasons, including taxation benefits, for them to set up shop here. These centres are contributing richly to stock of patents from India. This at least raises the worth of Indian researchers in the eyes of international investors, even if the direct benefits of R&D performed by them are minimal to local ecosystem.

MSMEs & Startups

Indian MSMEs contribute substantially to the GDP, but even more so to employment. Their R&D efforts are nominal as expected, but there are a few budding MSMEs that are quite active in R&D. This needs to be supported by the government and bigger R&D establishments. What kind of support and how to harness them is a question that can be debated.

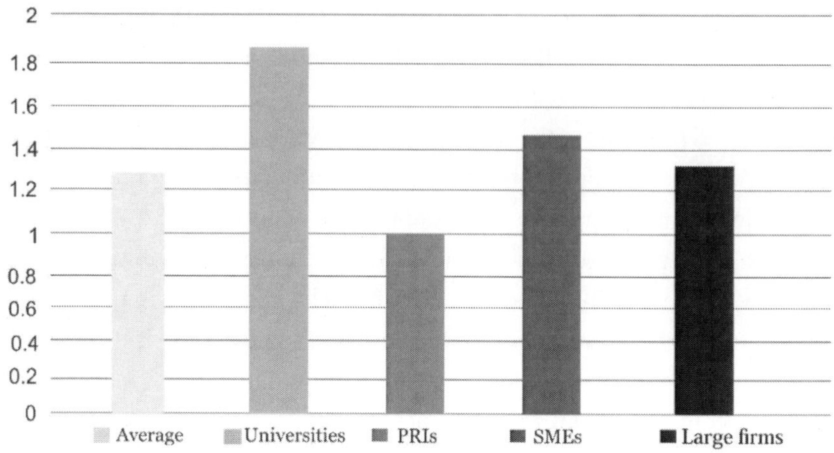

Figure 10. How different sectors (universities, MSMEs, large firms and public R&D institutions) in Korea contributed to patent-based productivity in a particular year.

Korea, Taiwan and Germany are good examples where MSMEs have been very active in innovation and contribute substantially to both GDP and employment. The Korean government has stipulated that all government funded institutions will spend 5% of their R&D budget in supporting R&D efforts of MSMEs [29]. Figure 10 shows how different segments of R&D institutions, manufacturers, universities etc. contributed to patents and productivity in Korea in a particular year [30]. It can be seen that the relative contribution of MSMEs and startups compared to big industries is quite good in Korea. This is not the case for Indian MSMEs.

The Indian government's efforts to reach out to MSMEs are disproportionately low both in terms of quantity and quality of effort, despite the presence of relevant institutional infrastructure [8]. If Government of India implements an equivalent scheme of mandatory 5% support from major R&D labs to these units, this will go a long way in improving their productivity (which is the lowest in comparison to that of all Asian industrial giants [18]), export worthiness and contributions to national efforts in manufacturing. A robust data with respect to their technology competence is not available. The existing data is mostly related to their financial worth or number of employees, range of products etc. Furthermore, heterogeneity and wide dispersal and often lack of registration with the government of these entities makes the response more difficult. Since the gap in knowledge base level (with respect to R&D institutes) is high, any initiative in this regard, coming from R&D institutes, has to be suitably adapted and stress more on practical technical training and demonstrations. This will have a high impact on productivity and have an inclusive effect as a large fraction of manpower (employees of these units) will be benefitted. These training courses should be short and must start with modern methods and techniques of doing things in labs or workshops, or it could be latest yet practical methods of quality monitoring and control or following international standards and best practices and sometimes can be more specifically sector oriented.

There should be a team of technology scouts sponsored by the government (detailed mechanism suggested later) to assess sector-wise the technology needs of MSMEs and the type of courses which can be useful for raising their productivity. These scouts should be working level personnel as the expectation is to collect ground level information. It is very important to seek feedback from participants on their technology needs so that courses could continually improve and orient suitably in content and style. Effectiveness of these programmes has to be monitored continually by technocrats. These programmes could be designed sector-wise and different national labs and institutes of higher education could be involved in the exercise. The end goal for these course providers is to target the improvement in productivity of these organisations. Online open courses could also supplement once the requirements are spelt out clearly. A very clear metrics should be evolved to monitor the progress in these directions.

A phenomenon worth noting and capitalising is that of startups, incubators and accelerators. Starting a new enterprise, based on innovative ideas, is becoming a very exciting venture for the young Indian S&T community, unlike earlier times when young entrants either wanted to go abroad or join some big lab. Old and large organisations, be they governmental or private, tend to become conservative in their attitude to being receptive to new and innovative ideas. So, these startups could unleash the creativity of young Indian scientists and technologists. In a recent survey, 25% of IIT Mumbai final year students expressed a wish to start an enterprise, spurning high salary jobs. This could be turning of the tide towards more innovation-based enterprises. As on July 1, 2018, there were more than 5200 startups in India [31]. These startups are presently employing about one hundred thousand employees and in all probability these are all high worth individuals. The government's launch of the mission 'Startup India' is again a worthy initiative, but there are few concerns like the machinery supposed to be implementing it, incomplete definition of startups, their linkage to innovation and hardware and how many manage to get funding and support. Regulations have been minimised in the new startup policy. They have decided not to define a startup more rigorously and they too want startups to go beyond the IT sector. Recently, the government has demanded job creation potential as an additional criterion. If implemented blindly, it could discourage many worthy candidates, as it is difficult to predict such possibilities right in the beginning. Thus, implementation of policy needs to be monitored carefully, lest it lands once again in bureaucratic spirals. The government has allocated a corpus of more than one billion dollars spread over four years. The fund of funds, which was expected to receive funds from Small Industries Development Bank of India, has received less than one tenth of the promised money and only one third of this has been disbursed to startups. It must be remembered that quick disbursal is crucial for a tech-based startup. If any credential verification is involved, that mechanism has to be made as smooth as possible. There are several other parallel steps that the government has taken to promote the phenomenon. It has set up a multi-tier network for stimulating startup activities, for example there are already 941 tinkering labs in schools, eight research parks, 11 technology business incubators, 50 bio incubators and 15 startup centres. [31] The existing startups have been able to attract a large amount of funding (rapidly changing figure to the tune of several billion dollars), though the total amount is slowing YOY basis, indicating the end of euphoric phase. But average funding per startup has gone up. It must be remembered that investors here are interested because of both local talent and the size of market in India. It needs to be deciphered how much is related to the size of market or genuine innovative talent. A question often asked is could Ola or Flipkart been as successful, had Amazon or Uber not been there on the scene? What would have been the response of funding agencies, regulators or even customers? The point is to highlight the existing prevalence of greater respect for international trends. Indian startups need to come out of this, be more original, rooted in the Indian soil of challenges.

Hardware related startups are relatively recent phenomenon, which is much more challenging, in view of lack of proper environment, both technically and financially. With startups like Flipkart, Snapdeal, Ola, Zomato and many others in the field, it becomes more challenging for hardware startups to attract funds because of their poorer gestation periods and lack of supporting technological infrastructure and supply of raw materials, components and prototyping facilities. What demands a careful study is from where these hardware startups originate, what type of institutions act as catalysts, what are the sectors where maximum interest exists and what are the type of hurdles these startups face. The answer to these questions may stimulate action to facilitate and accelerate the phenomenon of hardware startups and their contributions to the mission 'Made in India.'

A good number of these startups are based around IITs or IIMs and are geographically centred in Bengaluru, National Capital Region (NCR), Mumbai, Pune, Hyderabad and some other big cities. Those involved are students and alumni of these institutes with sprinkling of support from their faculty as well. More than one third of entrepreneurs are engineers and about one fifth of them have management degrees, few are PhD. In a welcome development, entrepreneurship culture in hi-tech areas has begun to penetrate beyond IITs.[10] These are good indicators but India needs to tap its vast reservoir of university students (and faculty too!). Another fact which has been pointed out by these entrepreneurs is that only a small

[10] A device to detect thinking patterns of the user (Mindbot), Tidal energy power generator ranging from five to fifty-kilowatt, 3 D printers of organic materials, devices to remotely monitor health parameters of patients employing internet of things are some of the ideas which have come from students beyond IIT's – mostly NIIT's. A Bengaluru Biotech startup has 3D printed a human liver and a device capable of both additive and subtractive manufacturing, built by young Indian students has got international recognition.

fraction of young engineering recruits from Indian institutes are worthy of working in these startups. The significance of quality of human resources cannot be overestimated. At the next phase of growth, input resources like availability of requisite facilities and human resources may become imperative. Besides universities, there are great possibilities of startups from national labs like departments of space, defence, agriculture, atomic energy and the like. The Department of Space has set a good example by encouraging such startups in the domain of satellite building and related applications.

In general, a small fraction of Indian startups are in high technology areas. Most of these are centred around subjects like analytics (450), artificial intelligence (200), internet of things (220), augmented reality (80), robotics (50), 3D printing (30) and blockchain (20). There is growth rate of about 30% in this sector of startups [31]. Hardware-related startups are a relatively recent phenomenon, which is much more challenging. What are the technological hurdles faced by hardware startups? Most of those in this business point out the transition from prototyping to manufacturing is the hardest part. There could be other difficulties of scaling up as there are not many experienced contract manufacturers in the Indian ecosystem. Some suggest to outsource these difficult parts from where it can come cheap. Shenzhen is a hardware hub in southern China. Getting printed circuit boards, or tool makers or precision machining of any kind is all well taken care of in this city. India lacks in many of these facilities, or otherwise these are scattered. There are some private sector firms supplying build-it-yourself kits for electronic devices. Intel, the semiconductor MNC, in its Bengaluru campus is facilitating the takeoff stage for these startups. The Intel India Maker Lab will offer startups and individuals access to development kits, reference boards and hardware and software tools to develop and test ideas. There are many such facilities springing up in different metro cities in response to these demands. In Bengaluru, a new maker lab has come up which is meant for hobbyists as well as industries. This has plans to upgrade their product range and facilities. Facilities like those in Shenzhen take time to build, but the government needs to be proactive in encouraging such facilities and maybe in some cases become a partner. This may set examples for private sector to take this up as well. Additive manufacturing needs to be exploited for such tasks. Such outlets need to spread all over the geographic regions in the country.[11] A recent development of various big industrial groups promoting startups from amidst their employees is also encouraging. The objective is to test the strength of new ideas and talent without getting their mainline industry bogged down. They support such activities by way of funding, retain some stakes and withdraw in case of failure. In case it is successful, they can leverage it for their mainline operations. Government R&D labs need to explore similar ideas but that may require a big change in government personnel policies. It is quite often that a product developed in a government lab, specifically for a project (but not linked to the main goal), loses its relevance close to maturity of the project. This is often sent to shelf or dustbin. Besides products, many useful ideas which are not of direct link to the mandate also face a similar fate. A government lab does not know what to do with such otherwise useful ideas and products. Such developed products, processes and ideas need to be taken to their logical conclusion. This process could be facilitated by encouraging young researchers to tap these ideas or half-baked products/ideas for setting up startup enterprises.

It is quite often that a suitable academic collaborator or business mentor can be a great catalyst for the process. Such collaborations between startup and national labs, universities and management institutions should be encouraged. Thus, services in kind can go a great way to compliment those in cash (grants) or tax rebates. Fast obsolescence is quite common in hi-tech industries. Indian national R&D labs and institutes of higher education which are best repositories of frontier knowledge base need linking up. The manpower should be ready to face technological challenges, assimilate new age skills and hop on to the latest ones. This is more likely to happen in the form of startups that are not bound by rigid mandates.

---

[11] Russia has recently come up with a prototype facility with very large number of fabrication facilities under a single roof. It is a big boon for young startups. India needs such facilities in large numbers.

As can be seen, there are plenty of research institutions and manufacturers that can link up and benefit all the stakeholders. India does have a lot of infrastructure in place, but what is lacking is quality, coherence, support from the right mix of policies and of course, the cultural gap. The next chapter focusses on some relevant policy issues.

Chapter 6

SCIENTIFIC RESEARCH AND MANUFACTURING: POLICIES AND ACTIONS

INTRODUCTION

No Science is immune to the infection of politics and corruption of power.

Jacob Bronowski

The progress of science, technology, manufacturing and their linkages in a country is dependent on complex factors related to historical background, traditional beliefs, spirit of enterprise among its population, type of government and policies pursued. This is despite the fact that the core of R&D is centred around individual enterprise. It is important to understand this in the context of a specific national situation, how these influences come about. Some of these remarks are confirmed from the way a number of Nobel Prizes in science have been bagged by a few countries (even universities and corporations), how world wars had a major influence on the developments of military technology, how cold war shaped R&D directions and many such others. How military technology later impinged positively on civilian life is a testimony to entrepreneur skills of the countries involved, besides of course the policies of the governments concerned. Choice of policies can strongly influence the progress as is evident from the achievements of countries like Japan, Korea, China and Israel in scientific research linked to manufacturing in the last few decades.

POLICIES THAT INFLUENCE GROWTH OF R&D

Which are the policies of significance to growth of R&D and of course its harness? It is not the policies directly linked to science and technology alone which affect growth, but many other facets of government actions that impact progress in science, technology and manufacturing. Questions are often asked about the Indian R&D sector and its relation to societal issues like growth of manufacturing and employment. There is a wide gap between expectations and yield from its R&D sector. It may require detailed examination to understand the several issues involved. R&D sector in India is so vast and heterogeneous that there need not be the same malady everywhere and consequently the same solution. There could be problems which require global attention and there also could be local level problems.

What is the purpose of R&D or technology development? There could be different perspectives and value systems of a scientist, technologist, science manager and policymaker at the national level. It could vary from a deeply altruistic one to an extreme commercial or populist focus. But it is a fact that the world over, there has been an increasing emphasis on more commercial exploitation of science. Internationally, the defence sector had been one of the major drivers of scientific research till the end of the cold war, but slowly civilian drivers have started becoming more dominant. This has influenced decisions on even basic sciences programmes as they too are getting influenced by private commercial sector. There is an increasing interest of private sector worldwide in basic sciences. The reason stems from the realisation among major private players about the long-term value of basic sciences, its potential to bring about innovations and generate related knowledge base. How is the situation in India and other developing countries and what set of policies and progress in science and technology are in sync simultaneously with specific requirements of developing societies as well as the international scenario.

One can identify all the players who contribute to scientific R&D, technology development and their deployment for manufacturing. Who are those who take up STEM R&D as a career and from where these workers are sourced? They come from universities and centres of learning within the country — some of those who decide to make R&D their career. In India, historically, there has been a segregation of roles of universities and national R&D institutions. By tradition, Indian universities have predominantly engaged themselves in teaching. Research in universities is predominantly related to pure sciences. STEM subjects' higher education in India is regulated by the University Grants Commission (UGC) and All India Council of Technical Education (AICTE). The government is mulling over some changes and there is a major committee looking into possible reforms. In most Indian universities, applied STEM research and particularly commercial R&D has been looked upon with disdain. This is in some ways similar to the situation in the early twentieth century America. There are few exceptions to this pattern in some of India's elite institutes. This way, the real potential of India's numerical strength continues to remain untapped, since a large fraction of STEM students are in universities and engineering colleges. It must be understood that education centres being fountainhead of manpower for R&D, education policies do influence the growth of R&D sector.

The manufacturing output is linked to R&D and technology development in terms of quality, quantity and type of products. As manufacturing grew out of traditional products (belonging to relatively mature technologies) to high technology products, it accelerated the demand for development of technology, helping establish a more intimate relationship between R&D and manufacturing industries. There are several nuances to hi-tech R&D and trade which are related to national prestige and development. There are issues related to dual use technologies and embargoes and restrictions of various kinds that inhibit growth of such sectors. Another major policy area that influences growth of R&D is related to immigration or loss of talent. There is a continual large-scale movement of skilled manpower, technologies and technological products and all of these are covered by policies on hi-tech trade, immigration as well as capability of institutions to attract and retain talent.

Each of the above mentioned policies related to higher education, S&T, manufacturing, international trade in hi-tech products and immigration need to be looked into constantly for their impact on growth of R&D and technology development and their consequential impact on hi-tech industrial manufacturing. These policies need to be dynamic as technology development grows very fast. In the following wok, each of these issues is discussed briefly.

More important than policies are their execution and action plans. At the ground level, there are a large number of agencies and ministries who are interacting with science, technology and manufacturing. The list is long, starting with ministries and departments of science & technology, human resource development, defence, DAE, space, agriculture, biotechnology, telecom, railways, heavy industries and commerce etc. There could be many more which could be responsible in peripheral ways. In such a situation, it can be imagined that having a comprehensive and holistic view for effective policy formulation, speedy actions and developing mechanisms for accountability can be quite complex. It could even delay the decision making processes in response to the fast changing technology scenario of today. Some significant facets of these policies with a few possible action points are discussed in the following work.

Science and Technology Policies

> For a successful technology, reality must take precedence over public relations, for Nature cannot be fooled.
>
> Richard Feynman

India as a nation has been enunciating its perspective about science and technology since independence in 1947. India's first policy resolution came in 1958, when the broad aim was "to cultivate and foster science research and temperament in the country." This seemed a statement from an amateur adolescent and dreamer, in light of low exposure, at that time, of the Indian public (as well as policymakers) to advances of science and technology and its potential impact on human welfare. The next science policy statement came in 1983, when India as a nation experienced its limitations for the first time in getting advanced technology expertise from the developed nations. This was highlighted in the backdrop of international sanctions after 1974 Pokhran nuclear tests. The next statement of intent in this context was in 2003, which again, it seems, was prompted by the technology sanction regime and related events following Pokhran II tests (1998). This was realised, by those who drafted this policy document, that advanced technology, which can withstand international sanctions, requires big investments and the need to stress on indigenous R&D and technology development in frontier areas. It also vaguely called for integrating national R&D missions with socio-economic sectors.

The dawn of 21st century witnessed rapid economic growth of many Asian nations exploiting S&T (along with trade and regulatory policies) and joining the international mainstream as well as manufacturing supply chains. So, the 2013 policy statement sensed internal lacunae in the above context, but again tried to point out the need for integration of science, technology and innovation for the social good. It also realised the limitations of public expenditure and expected much greater participation from the private sector in major R&D ventures. It also noticed with concern the necessity of attracting talent to science and technology streams, wooing them away from upcoming "glittering" sectors of IT and finance in the Indian jobs scenario.

India has witnessed frequent policy intents and it is not sure whether it has realised all these policy objectives. Any policy must be very clear and unambiguous and also must examine honestly whether the earlier policy was successful or not. It should resolve firmly to support some things and discourage others depending in what direction progress is expected. Does

India have in place the mechanism for harnessing whatever is already achieved so far? How fair and objective is appraisal of talent, projects and policy implementation? It is time these policies have fresh perspectives with focus on promoting linkages between scientific research and manufacturing. This may be one of the best meeting points of scientific research and societal needs. This is so as facing the challenge of disruptive technologies and probable consequent unemployment could become a major social problem. It is also required to examine the issues related to accountability on implementation aspects and how the complex machinery responsible for it is performing. An allegory below points out certain thoughts about policies, actions and accountability.

Mullah Nasiruddin and Donkey

> Mullah was a wise man but lazy. His wife used to nag him often for his laziness. "What is the worth of your wisdom if you cannot mitigate our poverty." He was used to listening to these barbs. One day he decided to visit the king of the land for a grant. He was well known throughout the kingdom and was well received in the king's court. The king asked him for the purpose of his visit. He explained, "Your majesty, I have come with a proposal for your kind consideration, if I am permitted to state."
>
> "Go ahead," said the king.
>
> "My lord, If I am allowed to have a donkey, I have ideas to transform it into a horse," said Mullah.
>
> "How it can be, how are you sure of your idea, why should we bet on it," the king demanded.
>
> "Sir, I promise to deliver it in ten years' time, and if I am not able to deliver, I shall offer my neck for it," Mullah claimed in confidence.
>
> The king got interested, he thought if Mullah succeeds, it will become possible to transform all donkeys of the country into horses and his army will become very powerful. "OK, I agree but your neck is definitely at stake, if you fail in the mission." The king ordered his minister to permit Mullah to take one donkey from the royal stable.
>
> Mullah walked out happily from the royal court. His friends, outside the durbar, were alarmed at his preposterous proposal and warned him, "What have you done, you have put your life at risk."

"Oh, there is no need to worry, in ten years' time, the donkey may not survive, or even me. The king may not be there for ten years, so keep cool," responded Mullah in his uniquely calm style.

The story tells a lot about how many of the R&D projects get sanctioned. There are many smart Aleks who make use of gullibility of their superiors. When projects are sanctioned, it is in the hope that they will be completed and the proposer is honest as well as competent to handle it. Here comes the need to verify antecedents (or track record) of the proposer. Also, most of the time, heads of institutes or those who sanction projects are rarely in power to see the fruition of these long-drawn projects and the next head is ill-informed to know the background and question them.

Also, the structure of government institutes is such that no action can be taken for failure to deliver and policymakers are led to believe that in R&D, projects can often fail. Most of the time, it is a bureaucratic or financial audit of the R&D project and not a performance-based audit. Accountability at all levels suffers.

Evaluating Role of R&D for Manufacturing in India

The policies enunciated in 2003 and 2013 are reasonably mature statements in some respects and have a lot of overlap. Both emphasise on the socio-economic relevance of science and technology. There is talk of evolving a comprehensive national innovation system. There is mention of evolving flexible mechanisms to help scientists and technologists transfer technologies developed by them for commercialisation and partner in the gains made. The real challenge is to translate these wishes into concrete actions on the ground. In pursuance of these goals, a bill on commercialisation of public-funded intellectual property was introduced in the Indian Parliament in 2008. This bill proposed generating an IPR-based publicly funded R&D and exploiting it commercially. This bill had some laudable objectives and was on the lines of celebrated Bayh Dole Act of USA. Bayh Dole Act passed in 1980 was a great success in terms of transferring technology from American universities to industries. Indian policymakers also thought of replicating those success stories. In case of India, it was not realised that there was effectively very little new technology knowledge base of this kind waiting to be tapped. Also, the industrial groups were not ready (or willing) to assimilate it. On the other hand, most of publicly funded R&D, constituting DRDO, DOS and DAE were not generating enough patents of commercial value. These departments get almost 70% of total government funding for R&D. CSIR, the main government industrial research body, has the largest number of patents among government labs and a very few of them have the potential to be commercialised[12]. In addition, Indian universities, which get about a paltry 4% of total R&D budget and are not geared up towards commercial R&D, have little to offer.

Most of the research and development inputs and outputs can be compared between departments and units and their impact on national industrial progress can be assessed qualitatively. While quantitative comparisons are complex, some broad assessments may be possible. In the present context, one can say that while India's strategic R&D could be claimed to have come of age with the help of sustained support over a long period of time, but a vast mass of its universities are yet to make a worthwhile mark[13]. Also, there is poor linkage between most R&D sectors and manufacturing industries. Evidently, the contrast between developing and developed countries is glaring in this context. The readiness of R&D ecosystem to face challenges of disruptive technologies is questionable.

---

[12] It has been observed in Indian context that wherever patents were encouraged or formed the basis of career buildup, it resulted in indiscriminate patenting. But there were very few licenses for sake of commercialization of these patents. CSIR is an example case. It costs Government exchequer two hundred thousand INR per Indian patent and two hundred thousand Dollars for American patents. Most of the time patent lapses after the inventor retires.

[13] There is strong argument to hand over funding to worthy hands. Some of the Indian strategic research labs have built sufficient infrastructure and human resources with sustained support from Government. They have developed unlimited appetite for funds. It is time increased attention is paid to Indian universities and other institutions deprived so far. A balance between 'strong & worthy' and 'weak & unworthy' needs to be restored otherwise the gap may widen further and large part of STEM workforce may go waste. 'Weak' may not always be unworthy.

Quantitative Inputs and Outputs for Scientific Research: General Comments

Whenever there is need to evaluate or compare R&D progress of groups, institutes or countries, a lot of care has to be exercised. Both R&D inputs and outputs need to be defined more quantitatively. R&D inputs are easier to define quantitatively in terms of funding and manpower, though even manpower is not equally salaried (and of equal worth) in different environments. As a major expenditure on R&D is related to manpower [8], these salaries may become an important aspect of inputs. Also, while comparing R&D expenditure as a fraction of GDP, there is need to consider the size of economy or its population[14]. In terms of purchasing power parity (PPP$), India's GERD approximated 50 billion dollars. India accounted for 2.7 % of the world's GERD. This amounted to 0.7% of its GDP [5]. Figure 11 shows how Indian expenditure on R&D as a fraction of GDP progressed over the years [5].

Nations like Israel and South Korea that have made maximum progress, were on the top with R&D accounting for about 4% of GDP. A more important point is that during times of economic crisis, most of the countries reduced their expenditure on R&D, except a few like Korea and France, showing their commitment to treating R&D as a key tool of economic/industrial progress. Another feature of R&D expenditure in India is that private sector contribution is also not showing any sign of progress.

Very recently, MNC R&D centres have started contributing heavily and if that is considered, then private fraction jumps from 30 to 38% [5]. Each of the earlier policies had suggested incentivising private sector expenditure on R&D, but with little impact so far. In India, there are several tax incentives which are offered and the government claims these are one of the best globally. DST white paper [32] describes the nature of these incentives in detail. There is need to examine the mode of incentives.

---

[14] For the year 2016 projected expenditures on scientific R&D were as follows: USA $ 514 billion, China $ 396 billion, Japan $ 166 billion, Germany $ 109 billion, South Korea $ 77 billion and India $ 71 billion. More important is how this research is made use of for growth in industrial manufacturing and consequent job creation.

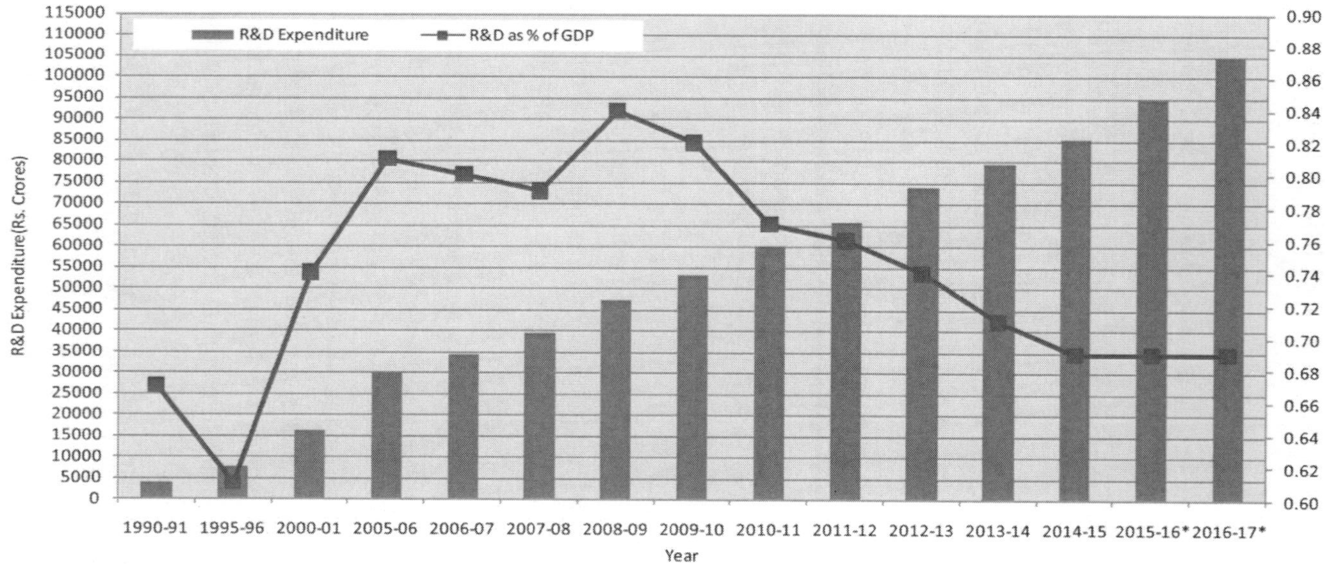

Figure 11. This shows the variation of total Indian R&D expenditure as percentage of GDP from year 1990 to 2016 (1crore equals 10 million).

Several governments prefer direct R&D project expenditure support instead of tax incentives. India needs to rethink its strategy of incentivisation as unfair practices in availing tax incentives by private industries could be derailing the good intentions.

R&D outputs are usually more difficult to quantify. These are often compared using a number of publications and their citations or a number of patents (and their commercial value) in case of technology development. India is making good progress in terms of publications as compared to the global average [5]. But a criterion which sometimes could be misleading is how much these publications are linked to the citations or to national industrial progress, even though they contribute mainly to the national prestige. US researchers are responsible for more than half of most cited 1% scientific publications. This has an indirect impact on developments in the high technology sector as well. The industrial sector in developed countries has the capability to assimilate and exploit such research results over a period of time, even high worth publications from developing countries are exploited by the same international industrial establishments. If one looks at countries like Korea, Japan, Israel and China, their patents are growing at one of the fastest rates and so is their manufacturing prowess. Though, in some cases industrial progress is less linked to patents and more to other factors like robust infrastructure and a favourable regulatory regime.

In countries like India, R&D is considered quite "cost effective" in terms of publications, mainly because of the low salary structure of its scientific workforce. But this differential with respect to western countries in wages encourages the best talent to migrate or move to other sectors of employment. It really needs a more detailed examination whether efficiency of its R&D, as per this metric, is the main contributory factor or the salary structure is the reason. For the real evaluation of their performance, these labs need to be compared to their counterparts in other similarly placed countries. India has a lot of lessons to learn from countries like Israel, Korea, China and Japan, besides some of the advanced countries like Germany and USA, for the sake of speeding of commercialisation of lab research. For example, Korean government labs have been playing a big role in promoting the growth of their private industrial giants like Samsung, LG and the like. All these countries have also facilitated the blooming of a large number of small startups as well as traditional MSME's that are known to be very productive. In India, government sector helping private sector tends to become a political issue and thus hampers progress. The Indian government can surely facilitate progress of MSMEs and startups. In India, the phenomenon of startups is more observable in the service sector, but not so much in hardware- related hi-tech manufacturing industries. The role of public sector R&D in contributing to this phenomenon has so far been limited and indirect at best[15].

The number of scientific or technical manpower per thousand of its population could be another yardstick to measure the intensity of R&D competitiveness generated within the country. The number of Indian scientific researchers is about 218 per million of its population as against more than 8000 in Israel and about 7000 in Korea [5]. This is related to its socio-economic conditions as well as traditions. India ranks even lower than some African countries in this respect. More than 95% of young Korean students are studying beyond graduation and one of the largest fractions is science/engineering based [33]. India does produce a large number of STEM graduates but they get dissipated, which explains the figures above.

Another criterion for gauging science progress is diffusion of knowledge, both scientific and technological. This diffusion is important across sectors and institutes. Internationally, knowledge is being jealously guarded nowadays. Earlier it was the strategic sector knowledge which was protected from theft or espionage, but now even knowledge close to being commercial (or potential to be commercial) is also not shared as easily as basic science. India's domestic performance does not inspire in this respect, for not necessarily the same reasons. Amazingly, most Indian scientific departments tend to work in silos when there is very little

---

15 Indian startups in the field of hardware and hi-tech industries lack support from its domestic industrial base like prototyping facilities, infrastructure, suitable mentoring, vendor base and marketing skills for novel hi-tech products. In most of these areas the public sector research labs can be of big help provided there is will.

knowledge base to be commercialised. There can be much greater synergy if they communicate and collaborate better.

It is said that basic science of today is technology of tomorrow — sometime this tomorrow can be fairly distant. Are basic and applied sciences progressing as parallel lines or there is hope for them to converge? Of course it depends on several factors. Basic sciences can play a significant role in nucleating innovations, training human resources and strategising for future which holds a big potential. India also has to realise the potential of diffusion of non-patent knowledge between institutes of research and from them to manufacturing industries.

As mentioned above, more than 70% of Indian R&D is driven by the government. The funding distribution pattern is not healthy from various points of view. It puts a lot of burden on public exchequer, unless it is a deliberate policy, as in the case of strategic sector, or when such investments continue to yield good returns. Also, government labs cannot be run as efficiently (not always true), they cannot attract and retain talent by offering international level of incentives and their linkages to industry are poor (even defence research mostly ends at government manufacturing for reasons of confinement to mandated output and at best resorting to exploitation of minor spinoffs if any. Even though government labs seem to be fulfilling their mandate, their efficiency cannot be gauged clearly in Comptroller & Auditor General (CAG — the national auditor) reports. An important point is that they are major expenditure makers of the nation on R&D and they are expected to be responsible beyond their specific mandate. In such a situation, all of them together should be covering most of the demands on the Indian science and technology scenario. Do they coordinate to that extent for the sake of responding to futuristic technology developments? Do they have a joint vision for the Indian science and technology scenario or they limit themselves to their own limited mandate?

As it may turn out, the contributions of many of Indian national labs to industrial progress has been at best indirect. For these labs to make a major impact on manufacturing sector and industrial progress of the nation, they have to diversify their interaction with industries and universities. Some of them do have some sort of mechanism in place for transfer of technology, but for genuine push to research-based manufacturing, that mechanism needs to be reinvigorated with more innovative methodologies as well as their scaling up. To encourage them in this direction, it is necessary that their appraisal, for the sake of provision of their annual funding or national ranking, should also factor in their capacity to stimulate private enterprise in manufacturing, besides of course indirect societal spinoffs. These will have several benefits like outsourcing of some of their problems and development of appropriate human resources beyond their boundaries, thus contributing to national goals and seeding for future entrepreneurship. This may necessitate a major review of functioning of many of these labs.

There is a talk of 'bureaucracy' in the report of the Prime Minister's Scientific Advisory Committee (SAC, erstwhile). This aspect should be understood clearly. What can be called bureaucracy in a scientific organisation? It is not the professional administrators, accountants or auditors who alone are responsible for this, but the scientific leadership and processes within that organisation itself. This is related to hindrances created in the path of genuine researchers and innovators because of reasons of perspective differences, personality issues, hierarchical issues, discipline, loyalty and other trivial reasons. Such issues are always brushed under the carpet and professional bureaucrats are blamed, though they have their own independent share. Any policy or strategy can be nullified by debilitating effect of bureaucracy of all kinds.

Policies must not restrict themselves with legislations and notifications, rather there should be clear-cut salient action points, monitoring mechanisms and the accountable agencies. In the following work, there are few action points about the policy issues mentioned earlier.

Beating Bureaucracy

First and foremost, the emphasis, for expecting any R&D establishment to respond to a challenge, needs to be on beating science bureaucracy and excessive hierarchy in government S&T institutions. These institutions cannot be run purely on administrative guidelines. The first step in this direction can be achieved by enhancing functional autonomy at every level. This will smoothen the impact of basic innovation processes, enhance accountability and speed up translating lab results to field/industry. Accessibility of youngest scientific worker to higher echelons needs to be increased. Many of these features are highly dependent on systemic defects but lack of effective leadership also plays a role.

Choice of Leadership

Careful choice of leadership in scientific departments/institutes will go a long way in improving the situation. Selection of leadership mechanism could be made more transparent and search committees should be publicised to the scientific community. This will ensure moral responsibility of search committee members. The present S&T leadership has to be held accountable in every respect. The level of support and autonomy has to be commensurate with performance. There could be several yardsticks for the choice of a leader, but the foremost is that genuine leadership has to respond to changing technology situations to manage its STEM talent efficiently. It has to be capable to attract talent, build up human resources, encourage innovation and make a dent in promoting linkages to hi-tech manufacturing.

National level Coordination

There is a strong need of active coordination efforts at the national level, of a large number and a variety of players involved in national innovation and R&D efforts. These players include government institutes, central, state and private universities, private research labs, private industries and innovation-based startups. Present day coordination agencies even at the highest level could be serving only limited strategic purposes. A national level mechanism needs to be evolved for all of them to interact with all those involved with the R&D business. This body has to not only monitor the status of S&T in the country, but identify gap areas and get action initiated, at public, private or collaborative level. This agency has to monitor progress (and build up rapport) of private sector labs, industries, MSMEs and startups as well. The first and foremost requirement for this is generation of a comprehensive S&T database. This agency could be aided by existing data collection mechanisms and other such institutions. The aim should be to enhance synergy, avoid duplication of efforts and monitor domestic transfer of technology (from R&D labs to industry) on sectoral basis. This must also monitor external balance of technology trade for the sake of maximising its benefits (by way of widespread deployment) and identification of weak points. This body should include representatives from private research institutions as well. It should be vested with authority as well as be held accountable. A good example for such a mechanism is that of Korea, where they have NIS which has been internationally benchmarked by OECD [34]. NIS could also address concerns of young innovators by acting as an interlocutor with policymakers. It could coordinate with several existing agencies, ministries and departments and their high-powered opinion may prevail over the others for the sake of speedy and right decision making. This could also address issues related to plagiarism, fake journals, credit hogging, promoting requisite culture and other issues related to science ethics in the country. (For details on this proposed NIS see conclusions).

Free Mobility of Manpower

Increased mobility and effective utilisation of expertise across the boundary lines of different R&D institutions (both public and private) is called for, otherwise their expertise is likely to remain underutilised. These exchanges should not be same as for the bureaucrats. This may require a suitable policy framework to be established. As at present, government labs helping private entities is some kind of a taboo and is looked upon with suspicion, though there are exceptions with high level approvals. Defence and pharma sectors are two worthy examples. This kind of cooperation and unfettered harness of human capital should spread to other domains and percolate to lower levels. Suitable norms and safeguards without restrictive bureaucracy could be incorporated. Even two-way entry and exit from universities or government institutes should become permissible for specific cases of setting up new science/technology-based enterprises.

Balancing Indigenous Technology Development and Imports

Experience has shown that forced external deprivation can be a blessing in disguise if internal support is provided to facilitate development of such technologies. Strategic departments like atomic energy, space and defence research are clear examples. These have achieved reasonable success in face of deprivation from hi-tech imports. Internationally, there were sanctions and these departments were forced to find ways and means (though time consuming) to solve their problems. This has lessons in general cases as well.

In a vast nation like India, import as well as development of hi-tech products continues in parallel while many a time, both developers and importers of technology seem blissfully ignorant of each other. There should be clear-cut policies in this regard, beyond the level of institutions. Such things happen because of poor coordination and left hand not knowing (or not willing to acknowledge) about activities of right hand. If such imports are a must, then there could be a better option to stop development work altogether and find alternative means, as it will save money and efforts. There cannot be universal rules and a case by case (or sectoral) evaluation by an empowered technical entity (in more significant cases NSTIS) is imperative. Technology considerations and long-term competency building should have a major say alongside commercial demands. This has certain riders, like in case of the strategic sector, where issues are complex. If a technology cannot be developed within a certain time frame, it is as good as not developed. Also, if a technology developed is not deployed at the right time and place, then it is as good as not there. Proper involvement of end users, for the sake of essential feedback, right from an early stage of development, can minimise rejection of indigenously developed technologies at a later stage after major investments of funds and human resources. Excessive respect for foreign hi-tech products needs to be curbed suitably. This has overtones of cultural legacy and must be handled accordingly. The best way to take care of such anomalous situations for a high powered agency like National Innovation System is to ensure that for each sector, a fair balance of technology trade is maintained. This task entails coordination between involved entities, incentivising technology development and promoting suitable tariff measures. Such a monitoring exercise by a technologically empowered agency could act as driver of hi-tech development.

Competition Imperative for High Technology

There are several lessons to be learnt from experiences related to Indian pharma companies and the most important is that having requisite human resources and fair competition can drive growth. Fear of impending competition from MNCs, post 2005, did trigger many of these indigenous pharma companies to spend more on R&D. Monopolies and 'orders on the platter' (as to PSUs or DPSUs) to manufacturers could discourage R&D and breed inefficiencies on the other hand. Thus, genuine and fair competition could incentivise innovation and it must be the norm.

Linking Scientific Publications to End Users

All Indian STEM subjects research work — publications, theses, dissertations or patents — should be categorised and graded in terms of having an end user (of any kind) in mind and declaration about the possible benefit in short, medium or long term. These features may be in addition to their international citations or commercial value. No dissertation or project report should be left unattended or unharnessed. It should be overseen by NIS. This will also give a national colour while making comparisons of individual scientists, projects or even institutions and injecing a dose of accountability. Even basic sciences could be categorised in these terms but not necessarily assessed like applied sciences and technology development. Basic sciences should continue to be valued for their potential to nucleate innovations, but totally useless R&D (with continued poor citation record and lack of linkage with any end-user group, even if academic) must be marked for discouragement. Those in basic sciences could be encouraged to develop human resources, transmission of knowledge base and join hands with technologists to strategize for future.

Mandatory Support to MSME and Agriculture Sectors

Agriculture is the backbone of most developing countries like India as it employs the largest fraction of their population. Similarly, MSMEs employ a very large fraction of Indian industrial workforce. Both these sectors are low in efficiency and productivity. Benefits of R&D labs must reach these sectors. All R&D institutions funded by the government should spare about 5-10% of their R&D budget, and more importantly, R&D efforts, for these sectors. This should not be in the form of grant (from these institutes to MSMEs/agriculture), but the required relevant R&D for them, imparting human resource training or transfer of knowledge base or upgrading their performance. This specialised outreach is besides transfer of technology (or spinoffs) which they presently pursue in the normal course. These supplementary efforts should be accounted for in their performance appraisal in addition to their basic mandate. Efforts of these labs should be directed towards improving productivity/efficiency of these sectors. There are departments that are already working for these goals, but other departments will be induced to reorient some of their R&D directions and be held accountable on these counts. There should be broad guidelines and there should be freedom and flexibility in fine tuning the directions of such R&D.

STEM HE Policies and Made in India

> The only thing that interferes with my learning is education.
> Albert Einstein

STEM HE can be termed the fountainhead of all scientific research, technology development and industrial manufacturing. All R&D institutes and manufacturing industries source their manpower from Indian universities (both public and private), NIITs and IITs. This is equally true for strategic sector institutions and public sector defence manufacturers. There are several streams, sub-streams and levels of higher education in STEM subjects which are needed by industries and institutes of higher education. Similarly, there are several related skills required. For each of these streams there could be continual change in demand, while supply side does not change so fast. In addition to these problems are the issues related to quality of manpower. So, despite India producing large numbers of human resources in each category, problems exist. This is one of the serious challenges faced by Indian policymakers.

Challenges in STEM HE

Latest figures indicate that India has 864 universities which comprise among others 44 central universities, 540 state universities, 122 deemed universities, 90 private universities and 75 national institutes of importance that include Indian Institutes of Technology (IITs), All India Institutes of Medical Sciences (AIIMS) etc. India has the capacity to produce about five lakh engineers, more than ten lakh science graduates and about four lakh engineering diploma holders from polytechnics. There has been a sharp increase in both the overall number of students and proportion of young people in higher education. About 25% of the graduates now go in for higher education. At graduation level, about 20% go for science subjects and about 5-10% for engineering. All IITs together churn out about 10000 engineering graduates and about 200000 graduates come out from second and third rung institutes every year [8]. It must be remembered that the IITs, which have the brand value of high international standards, admit only 3% of applicants. From among these IIT graduates, a fraction either goes abroad (a declining trend presently) or gets involved in activities not directly related to their subject. In 2015, the Indian scientific workforce was about 0.5 million, out of which less than 300000 were involved in active research, while the rest were contributing to auxiliary or non-technical support [5]. There were more than 15000 doctorates belonging to science and technology subjects. Most of these doctorates are from very few institutes. Despite such large and diversified STEM workforce, Indian industries, particularly young startups, complain of inadequacy of fresh recruits, with most startups requiring recruits molded for innovation-based activities rather than routine tasks. There appears to be a divergence between marketplace needs and qualifications and orientation of fresh graduates, postgraduates, doctorates and diploma holders. A very large fraction of these STEM graduates has to shift their career track as they cannot be absorbed suitably. The most important reason for this is that a major fraction of Indian public sector R&D institutions and university R&D is devoted more to strategic sector or basic sciences and not directly concerned with industrial growth. Even teaching jobs in universities cannot absorb most of them because of poor quality control of doctorates generated. Specialised and highly qualified workforce do not find the right placements in the industrial sector too, for reasons of mismatch of interests.

Indian STEM HE is confronted with an eternal question about match of supply and demand. Addressing both quality as well as quantity issues simultaneously in demand of appropriately skilled and educated workforce is the biggest challenge faced by the Indian education system. Demand pressure on trained STEM workforce has come about mainly with the rise of IT sector. New recruitment actors on the STEM scene are IT majors and MNC R&D centres. Appropriately qualified STEM HE manpower is desired at an accelerated pace for multiple reasons. An associated important prerequisite is attracting and retaining students in STEM streams. Similarly, on skills front, there is problem of the stigma associated with vocational education stream. There are problems related at both supply as well as demand ends. The most significant challenge is related to attitudes and mindsets and the inability of the current ecosystem to value and manage its own talent. This is also intimately related to paucity of resources for higher education in mass education institutes and on top of it funding being offered to these institutions is also not being used efficiently.

The first and foremost issue in managing these institutions well is about funding. Allocation of funds to higher education in India was on constant growth path since independence in 1947 from about 0.19% of GDP to about 1% by 1980, after which it started declining and continues to this day, even though in absolute terms it may be increasing. In 1997, the Ministry of Finance declared higher education to be beyond subsidy, labelling it as "merit 2 good" [35]. Central universities are reasonably funded but state universities are highly impoverished and need funds for bare sustenance. This led to introduction of Rashtriya Ucchtar Shiksha Abhiyan (a mission for higher education in Indian languages) to support such universities. These universities are considered educational institutes and are funded under the head of education for their infrastructure requirements through the Ministry of Human Resource Development. DST, CSIR and UGC are the main support agencies of these universities for R&D projects. Some of their projects are also funded by agencies like Board of Research in Nuclear Sciences (BRNS) and DRDO. Recent intentions or declarations of major amounts of support for institutes of higher education can only be considered too little, too late. Neglect of decades cannot be compensated in a short time. This level of support is quite nominal when compared to what is actually required, as these universities constitute a major fraction of enrolment in India. In such a situation they cannot be expected to contribute the quality manpower required for R&D and related manufacturing.

The whole situation has to be viewed in a much broader context of globalisation of higher education. There are a large number of students in India as well as many other developing countries who migrate to developed countries for higher education. Many of the universities in these developed countries are sustained by tuition fees from these international students. This amounts to these countries gaining billions of dollars from developing nations and the amount continues to grow. Many developed countries find it difficult to sustain this demand within their shores and are planning to set up offshore branches in several countries. In future, there is likely to be increased pressure on WTO to treat higher education as a service, which developing countries may find difficult to resist for long. India and other developing countries need to wake up to such stark realities. The rationale for treating higher education as a "commodity" is that it leads to higher income for individuals benefitting from it. This may be partly true. STEM HE needs to be considered in a different light as it goes beyond individual incomes and also contributes to building up of developing nations by way of sustaining activities which are strategic, and thus should be considered as "public good" rather than "tradeable good." STEM HE is the biggest pillar supporting even strategic sector R&D and manufacturing. Accordingly, it must be viewed as segregated from the rest of higher education. This must be funded, managed and monitored accordingly.

There is need to understand the flow of funding for R&D in universities, its influence on the choices in directions of research, how various funding agencies operate in sanctioning and reviewing the progress of these projects and how this influences manufacturing sector industries. Presently, most of this is left unmonitored (or poorly monitored) and precedents, legacy and limitations of infrastructure dictate R&D directions in universities. This is one of the major causes that contributes to lack of coherence and connect between scientific research in universities and industrial growth. For example, DST, which is the largest funding agency for extramural projects, has a system of programme advisory committees to consider, sanction and review projects from most higher education institutes. They have a role of simply cross checking the scientific goodness and viability of the projects/proposals reaching them. These committees are not always proactive to the extent of influencing a detailed course of R&D directions, promoting competency building, mentoring for the sake of addressing gap areas, orientation to industry requirements etc. Same is true of many other funding agencies. Evidently, the increased quantum of funding to university STEM departments together with its effective management can make good impact on R&D for industrial manufacturing.

Elite versus Mass Education

Several studies worldwide have pointed out how in the new age economy, higher education and advanced skills play a big role in making a nation's labour force competitive. The manpower in such nations is less prone to unemployment and gets better wages. There is regular talk of the rising tide of technological/technical upgradation in BRICS economies. In this context, a group of American researchers evaluated threat perceptions to US position vis-a-vis BRICS [36]. Their findings make interesting reading and are a lesson for Indian policymakers. Their observations point out about the education scene in BRICS countries. "Their mass institutions are receiving fewer resources," the study says, and that's where most of the students go. In 2009, 2.1 million of 2.5 million bachelor's graduates in China matriculated from mass institutions. In India, it was 2.2 millions of 2.3 million. It further points out, "There is widening funding gap between top schools and mass institutions and this has broad implications." The gap has the potential to slow economic growth domestically, deepen income inequality and create less social mobility. The significance of higher education (beyond graduation), particularly in the context of mass institutions, is yet to sink in the psyche of public as well as policymakers. A very small fraction of private sector is willing to invest in this sector for their own reasons (recent effort to rope them as institutions of eminence could solicit few worthwhile proposals). The constant talk of having world-class institutions and banking on international rankings may push developing countries to be elitist. This has to be viewed together with the fact that socio-economic conditions in developing countries influence less number of individuals going for tertiary education. It is noteworthy that in Korea, about 98% of students go for higher education and in India, the number is about 24%. It is also obvious that elite institutions in developing countries with limited resources cannot cater to such a large fraction going for higher education.

What happens as per their observations is, "BRICS countries are doing a good job of educating students at the elite levels, but they are not doing a good job of educating much larger number of students at the non-elite levels who are also fundamental for the economy." In their view, the engine of these new economies is the rest of the population — people who attend mass institutions. These observations are very pertinent in the case of STEM subjects. This has to be viewed together with the fact that a good majority of graduates from elite institutes tend to migrate out of the STEM stream or country. Finally, the report points out, "If these economies cannot cater to this mass segment then they can never be able to compete with western scientific-industrial complex." This is because western countries have been able to organise their higher education (particularly in STEM subjects) system such that these second-tier institutions are well taken care of, as they call them "guts of the system." One of their major recommendation worthy of consideration for such economies (more so to India) is to increase graduate education and funding in R&D. This must be done in the context of the overall economy, population as well as GDP.

Providing quality higher education to a large number of poor aspirants is a tremendous challenge to the Indian state. But it must be realised that this is an unmissable opportunity too. This is no less significant than having good infrastructure. The problem is further aggravated by the fact that the Indian ecosystem finds it difficult to retain all of them and a large number of them migrate in the absence of requisite quality of jobs. Thus, most of such financing of tertiary education or research fellowships serves as a suboptimal temporary employment scheme. The nation does not benefit from their expertise. It is imperative to efficiently match supply and demand of such trained manpower and ensure such education is seamlessly linked to fruitful employment of these high worth individuals. This can happen if all STEM HE fellowships are sanctioned with some academic or industrial or governmental user in sight. The competition spirit and system of incentives and disincentives should pervade the methodology of sanction, review and ensure employment. There are additional constraints like being responsive to affirmative action and yet ensure fair competitive spirit and due respect for merit. Another issue often overlooked is that affirmative action is not purely for political or egalitarian reasons, but is also an opportunity to widen the search base for talent. This also calls for increased competition even within students availing affirmative action benefits. A noticeable point is that such talent may be the least volatile in the Indian STEM ecosystem.

A very pertinent and important challenge is to make the whole system of financing higher education self-sustaining. It has to be understood that the government may like to gradually limit support in some ways and it could trigger protests from aspirants. This may call for the government to evolve a suitable policy that balances constraints of finance to satisfy demand on one hand and ensure fair competition and affirmative action on the other. This may require all applicants to be divided into different categories based on economic potential, merit, aptitude and belonging to category of affirmative action or not. All higher education institutions require an optimal fee structure for their running expenses. Additionally, a large number of STEM HE institutions have to be given autonomy to earn from their joint ventures and consultancies with private industries or other such institutions. Such measures will minimise the load of fees on students/researchers. All deserving students selected on the above basis have to be compensated with scholarships and soft loans. To keep the competitive spirit alive with the help of a system of incentives and disincentives, these scholarships and loans could be at differential rates and on returnable basis depending on performance of students/researchers. All those within the affirmative action group should also compete between themselves for differential rate of scholarships. Growth of private sector in higher education should get linked to proceeds from harness of the knowledge base developed by them rather than purely on fees from students. The government has substantially raised the amount of fellowships for certain category of researchers. Stipend amount is important to a certain extent, but ensuring all other conditions, like capacity to absorb them usefully, are crucial for evolution of appropriate ecosystem. Recently, the government has expressed its intent to set up several more IITs in the country. This may be a move in the direction for ensuring high quality tertiary STEM education for a larger mass of population. But it has hitches related to availability of the right kind of faculty for all these IITs. This calls for incentivising recruitment of the same. Several other such hurdles have to be overcome for attaining the goal of massification of tertiary education in STEM subjects.

Role of Public and Private Sectors in STEM HE

Private spending both on STEM higher education and R&D has increased in recent times in most developed economies (OECD Report) [37]. Internationally, private sector is now also trying to influence directions of research, including basic research. This indicates a growing linkage and overlap of roles between private and public sectors internationally. Increase in private spending on education and basic sciences has come about because of the realisation about the significance of innovation in generating new products and processes and the potential role that basic sciences can play. This is not yet realised by private sector educationists and industrialists in developing countries, including India. What policies can catalyse this process?

Traditionally in India, charitable organisations and missionaries have played a great role in building educational institutions, including some in higher education. But in recent times, this source of funding is transforming to commercial ventures. Exceptions are a few with nominal impact. The number of institutions and the courses offered by them has increased rapidly and they are not monitored very efficiently, both for their quantity and quality. In view of non-availability of sufficient funding from the government, it is imperative that the private sector is involved in innovative ways. One of the approaches is to invoke the corporate social responsibility (CSR) mechanism (a mechanism which is mandatory for large corporate houses —to spend a certain fraction of their turnover on some social cause). A recent report on the review of status of education in India by Confederation of Indian Industries [38] noted that only 27% of companies (part of their survey) were meeting the proposed norms of corporate social responsibility spending. It pointed out that a significant potential exists in this area for generating additional resources from CSR funds of Indian companies. The report said, "While the intent for using their surplus funds for improving the education sector in the country by Indian companies is evident, what may be required is a clear and unambiguous policy from the central and state governments to leverage this funding in an effective manner." It also commented, "While almost all major committees and policy documents have accepted the need for increased involvement of private sector in higher education, there is also lack of clarity on funding pattern, incentives and regulatory oversight." There are several kinds of regulatory excesses which discourage private sector. Government policymakers seem to be lacking in imagination in making use of this willingness. In case of private sector contributions, autonomy with accountability is a desirable principle to be followed. But in garb of autonomy, a large number of such private institutions indulge in profiteering and siphoning off large fee amounts and funds and grants taken from the government. Whatever solution is found to this problem, due care must also be taken that genuine talent from poor background does not suffer in the process. For the sake of a few black sheep, all private institutions cannot be allowed to suffer as this affects the health of the overall STEM HE ecosystem. This may suggest a middle path, offering some sort of graded autonomy for a limited period. STEM departments could be graded in say a three tier structure with different levels of autonomy. Submission of all academic and administrative performance details should be made mandatory, based on which they should be ranked on the national level and graded for the purpose of identifying the level of autonomy entitlement. Each year they could be upgraded or downgraded depending upon performance. The first and foremost step is about academic autonomy. This allows these institutions to collaborate with other institutions or industries. Their main capital is knowledge and they should be allowed to sell it and earn their upkeep. Such earnings should become a norm. Both their faculty and researchers/students can participate in this process. There are several ways this can be capitalised. Good performance on this front should be rewarded by offering the freedom to choose curriculum and R&D directions (it could be in sync with those industries) and other types of autonomy. This autonomy may also include unhindered two-way flow of human resources from universities to industry for the sake of transfer of knowledge base.

It is quite surprising that no Indian institute figures in a compilation by Times higher education study, which ranks institutes and universities based on their linkage with industry [39]. Having such linkages is crucial for developing countries for the sake of growth of R&D culture in industries, reinvigorating university science system and also facilitating retention of talent within STEM streams as well as within the country. In most developed countries, industries and education system have both benefited each other immensely and industries are planning even deeper forays into the territory of education for furthering their own goals. POSCO, a Korean iron and steel company, is an example worth notice. It has set up a world class university attracting reputed international faculty. It has also benefitted from the knowledge base generated. Industries in developing countries should learn lessons from this. There could be several examples but that of Korea stands out where patent productivity of universities compares favourably even with respect to government national labs. In Korea, research-based ventures have originated from university researchers and professors and it compares quite favourably with respect to the overall national performance. In India, only a handful of universities display such performance.

Defence labs have recently sought services of few IITs for development of specialised missiles and planes. They have brought these IIT departments together for this purpose and allocated an order of a billion INR for this purpose. Such initiatives must go beyond defence-related work to many other sectors like railways, telecom, agriculture, renewables, energy storage, transportation, urban revival etc. that are involved in major expenditures or imports of futuristic technologies. Making use of expertise in institutes of higher education calls for formulating such requirements into a scientific or technical problem at the user's end to be followed by scouting for relevant expertise in universities and institutions of higher education and coopting them and finally coordinating the whole exercise. This problem solving should cover as many rungs of institutions and fresh startups as possible. This may stimulate strengthening of the ecosystem. This may impact employment possibilities for appropriate workforce from STEM HEIs with increased overlap of interest of universities, industries and technology-related government departments[16]. For accelerating and broadbasing the process of university-industry interaction, it may be worthwhile to involve a large number of MSMEs that are presently employing the lower end of technology. Universities can also contribute to solving several R&D-related problems faced by them. Korea has exploited this route with great advantage by creating a university-industry consortia with MSMEs. In India, the potential for such collaborations is huge.

---

16 DRDO reposing faith in consortium of labs from IIT's is a big leap. Instead of giving small projects to individual departments doling out funds in small tranches, it is much better to give them collectively a bigger responsibility. This will help these departments build appropriate infrastructure, expose researchers of these labs to perspective plans and problems and help reorienting thought process of these researchers. More of such big national labs have to think and act like that. This may help evolve fresh perspectives and better linkage may happen between type and strength of R&D manpower and projects of national significance.

For scientific research in universities to contribute to industrial manufacturing, all three stakeholders such as universities, industries and researchers have to reorient themselves. Aiming to bring universities and industries together calls attention to the problem of attracting young people to the careers in science and engineering. This will have serious implications on availability of researchers. It was reported some time back from an India wide survey that in VI standard, about 22% students were motivated to take up science as a career, but by the time they reached XII standard, only 6% were motivated enough [8]. This has at least two main inferences: that the quality of Indian science teachers is not good enough to motivate them and these students learn about lack of reassurances on careers related to science within the country, as they grow up. These figures may vary depending on the subject. About providing reassurances, there needs to be better coordination between recruiting institutes and industries on one hand and colleges and universities on the other.

There are several modes in which university-industry interactions can fructify. While industries in general could pose research problems to universities, the latter could also give leads to industries. The example of modes of collaboration between Korean universities and industries, as shown in Figure 12, shows the various modes in which universities in Korea have collaborated with industries, also pointing weightage for each mode[17] [30]. There could be variations depending on the country. Complete focus on basic sciences in Indian universities is utopian in nature as this will deprive them of funding support from industries and wean researchers away from national missions of social/industrial relevance and aggravate problems related to their employment. The Korean government did dilute laws on autonomy of its universities to facilitate two-way traffic of human capital between universities and industries, for example in the context of their Innopolis (or innovation cities).

Researchers are the key to the whole process. How are universities and researchers inhibited in contributing to industrial progress? How university

[17] Korea has been fairly successful in generating partnerships between universities and industries. Most important requirement is simplicity of rules and bureaucratic procedures for two-way movement of human resources. Most of their Innopolis or Innovation cities progressed based on such steps.

researchers make choices in their R&D directions, how they carry it out and the results of research are passed on to industries or startups? Indian university researchers presently are remotely placed from such scenarios.

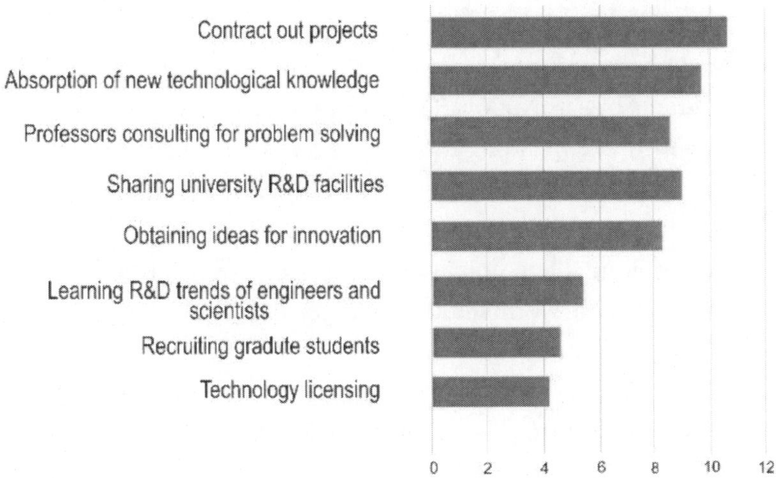

Figure 12. Various modes of collaboration between universities and industry researchers in Korea indicating weightages for different modes.

An important example worth mentioning here is that of American enterprise management groups guiding American researchers/professors for commercial exploitation of their innovative research. The programme called Innovation-Corps started from a programme called Lean Launchpad at Stanford University. The programme is meant to educate professors and academicians about how to evaluate their research against customers in the market. As a pilot study, it formed 21 teams of three members each. Each team consisted of one professor/researcher (principle investigator), one graduate student and one mentor, usually an experienced entrepreneur. There are more than 300 such teams from about hundred universities in America. Their experience of the exercise is quite revealing and may be applicable in parts to a large number of university researchers in general.

It was found that most professors were not venturing into commercial sector for reasons relating to the fear of failure, fear of cold response from customers and inertia to deviate from their profession. (In India, there could be additional fears about high cost of capital, getting clearances and taxation and even closing the company in case of failure.). Most of the research teams felt much better after this experience and realised that customer feedback could influence the course of their research work also in a positive way by upgrading their developed products to be useful and suit the market requirements. Most of these teams were for projects based on research grants by US National Science Foundation. While in USA, this project was aimed at expanding the base of research being made available to industry, in India (and many developing countries), which has a large number of universities and excellent management institutes, this could be a good starter. Making them join hands for the sake of fructification of scientific research is a worthwhile challenge. India may consider having such programmes on a national scale for at least projects funded by DST, CSIR, UGC etc. Presently, such projects may not be in large numbers but if universities implement reorientation of their R&D directions, then more and more of such projects will become eligible for such monitoring/ management support. This should unleash huge creative energies channelising them in the right directions. Normally, programmes funded by DAE, DRDO or DOS have well defined end user goals but, in those cases also, spinoffs are often frittered away. These programmes can also be suitably covered under this scheme. The aim is to move the academics making progress to commercialise their R&D as efficiently as possible. The advantage of the sheer number of researchers in India will show up in this process. In this connection, not discouraging (preferably encouraging!) two-way traffic of manpower will prove quite useful, where "failed" entrepreneurs can return to their original lab.

Managing STEM Talent in Universities

All developed economies are fiercely in race for attracting and retaining science and technology talent. It is based on the premise that their state of advanced development with respect to developing economies is mainly function of their superior technology and talent. With advantage of demography shifting in favor of many emerging economies, they do feel threatened. To keep that technology gap alive, they have to wean away talent from these countries. Their policies related to immigration, student scholarships, social security net for these technocrats and employment policies have these objectives in mind. Earlier outsourcing and now global in-house centres are steps in the same direction. Internationally, in the last three decades, there has been a fivefold increase in migrants from many developing countries to developed nations [40]. In India originally, the trend for such migration was prompted by higher education, but soon transformed to education followed by jobs and permanent settlement. It is noticeable that all recipient countries are well developed technologically and have a robust education system and linkages between university labs and industry. Students make selection of their target country based on the reputation of the institution or programme, recognition of the programme worldwide and prospects of getting a job in the host country and sometimes the language of communication [40].

The ultimate yardstick on the quality of an institution in India should be its capability to attract and retain talent. The First requirement to retain talent is that it needs to be endowed suitably, sometimes to the extent of pampering. Bureaucracy of any kind should not be allowed to fritter it away for petty rules and regulations. There are lessons from the way MNCs are tapping Indian STEM talent for a vast range of requirements for their global operations. India as a country is opening up in several ways and this may be the right time to bring back a large number of talented Indian migrants. The wave of protectionism sweeping several nations may offer attractive possibilities to many developing countries to bring back talent. The present programmes of different type of seemingly attractive fellowships for Non-Resident Indians (NRIs) are insufficient. These may suit needs of a few individuals, but not what is required at the national level. Individual institutes or state governments initiating such efforts may not be very effective. A high-level national recruitment board may be necessary. This could be on the lines of the very successful Thousand Talent and Ten Thousand Talent programmes of China. It involves high level, multi-tier selection process by scientists/technologists, free of any bureaucracy. These talented individuals should be involved in projects or universities/HEI. Depending on the level of recruit, the selection machinery could be different. A high-level selection machinery will be able to place them suitably (based on their expertise), offering them due incentives and toning down "resistance" of existing workforce in those areas. They could be deputed to solve pending socio-technological challenges being faced and augmenting and training the human resources. Depending on the level of recruit or impact making innovative idea he/she brings in, he/she could be encouraged to even set up sector specific greenfield institutions. Another initiative in this connection is about having semi-formal virtual networking of Indian diaspora and Indian university STEM workforce. This may be for those migrants who are not able to come back. This could open more possibilities of joint projects, collaborations or even post-docs for young researchers, or refereeing the project proposals or even individual appraisals.

Following are some practical action points in pursuit of requisite policies which can influence STEM education for the sake of linking R&D to manufacturing sector.

Redefining National Rankings

The new national ranking of Indian higher education and research institutes has been compiled by the Ministry of Human Resource Development. Its third edition has been announced recently. These rankings have slightly altered the weightage criteria in line with international rankings. This is a step in the right direction, but the methodology has to evolve further and some modifications in weightage factors of parameters involved is called for by incorporating local situations. For example, some of the additional parameters which need to be incorporated could be related to capability of institutions to attract and retain talent and linkage of institutions to industries. These are parameters on which India as a nation is poor and there is need to encourage efforts in these directions. There could be more such parameters which need to be considered. Some of these changes in weightages at macroscopic level may be less than effective. For example, more important than weightage for

R&D is the type of research universities carry out and its linkage with application, say industrial manufacturing or setting up technology-based enterprises and contributions to building national industrial base. Instead of limiting to universities, these ranking exercises need to be extended to individual STEM departments as there is a lot of variation in performance of different departments within the same university. This will be useful for all relevant stakeholders like aspiring students or recruiting industries.

Segregation of STEM HE and Graded Autonomy

If India aims to tap STEM departments in universities for promoting industrial growth like in many developed countries, then few radical steps need to be taken. STEM departments in universities have to be treated differently from other subjects like humanities. This is for the sake of few specific objectives like organising targeted support and making them specifically accountable for the same and defining entitlement to level of autonomy to collaborate with industries or other institutions.

Matching Supply and Demand and Channelisation of Human Resources

Efficient and flexible generation of STEM workforce can go a long way in reducing the problem of unemployment (or under employment) of STEM workers and facilitate retention of talent. This calls for monitoring performance, proper aptitude tests at every level and suitable mentoring to guide and channel young students and researchers into specific career or research directions. The careers could be categorised into teaching in schools and colleges, industrial production, services, maintenance, engineering or research and development. Even financial support could be decided more efficiently based on parameters of candidates like his economic situation, merit, aptitude category and affirmative action requirement, as mentioned earlier. Students joining the stream as per their aptitude may be offered preferential financial support from the government. This could be by adjusting scholarship amounts and interest rates on student loans. This could be further supported by measures like exit tests, where those who fail have to return the financial support already provided. Such measures may promote healthy competition for all and such channelisation based on aptitude may allow the government to fine tune supply and demand together, of course with the help of database from potential employers, both industrial and academic.

Autonomy and Accountability of STEM HE Institutions

Innovative solutions have to be found for giving autonomy to STEM HE departments in universities and simultaneously preventing harms of deregulation. The first demand on the institutions is that STEM HE in colleges and universities beyond the level of graduation must be linked to R&D. They must get a message that better R&D with links to end-user/industry/enterprises will fetch better autonomy and support from the government. Autonomy could be provided in a stepwise manner. There could be several gradations of autonomy such as freedom to launch collaboration with industries, allowing industry friendly changes in curriculum and directions of research, free movement of human resources with collaborating partner etc. This could also mean giving freedom to students, researchers as well as professors to offer consultancies and joint projects. This autonomy could be withdrawn or upgraded depending upon how it is used. Thus, autonomy with accountability needs to be reinforced. STEM institutions should be incentivised to tune to revenue based on knowledge base generation rather than student fees and government support. The fee structure should be considered together with the availability of scholarships and easy loans. A liberal atmosphere on this front could permit higher fee. This may vary from one institute to other. These rules need not be universal, once a detailed database about performance of these institutions is available.

Affirmative Action

Affirmative action for STEM students/researchers should not be treated purely as a political issue with egalitarianism in mind. This should be treated as an opportunity for widening talent search in deprived communities, including women. This can be facilitated by having competition within each such subgroup by having a differential rate of scholarships. This will ensure incentives to more talented among this group.

Teachers as Pivots

All STEM teachers both at school and college levels need to be incentivised in terms of their service conditions. How is the job of a teacher, in terms of service conditions, as compared to other jobs in the country? Making a break from the past is possible only with the help of deep pocket autonomous private sector industries venturing into STEM HE that are not bound by the legacy of earlier service conditions. These should be able to attract quality faculty and have objectives to earn revenue from the knowledge base generated, rather than student fee alone. Teachers at all levels and in all sectors have to play a pivotal role in building up STEM HE, research and even manufacturing ecosystem. If talent cannot be attracted to this stream, it may impact a large number of research programmes and development of high technology in the country. Selection of teachers, curricula and system of examinations should be able to wean students away from the bane of coaching classes, for the sake of promoting innovation.

Fellowships and Linkage to Employment

The Government of India has recently announced a scheme for large value fellowships for STEM researchers from few eminent institutions. This is aimed at attracting and retaining talent in STEM subjects within the country. This step may or may not serve the purpose. The reasons being that is the direction of research may not offer satisfactory employment opportunities to researchers within the country. Unless the choice of research direction is carefully made in consultation with prospective stakeholders within the country, the scheme may not provide much hope. In general, it is desirable to follow such guidelines for all research fellowships and these should not be allowed to degenerate as a launching pad for moving out of STEM ecosystem or the country itself.

Balancing World Class and Mass Education

There is a big race to develop world-class institutions in many developing countries. India is also making progress in this direction. While this goal may be laudable in itself, it does have some disadvantages. India has already experienced this as a large number of its bright students trained at IITs on government expense have been lost to developed countries. Also, it must be remembered that talent does exist in many mass institutions too, but has less chance to progress because of lack of opportunities, infrastructure or poor guidance and mentoring. It is thus imperative for countries like India to strike a balance between supporting world class and mass institutions of STEM HE.

Bridges with Polytechnics

There is need to inculcate among STEM students, interests in experimental sciences and do it yourself (DIY) culture. Presently, in small number of schools, Atal innovation centres have been launched with a similar purpose in mind. These are centred around a few contemporary subject areas with a small number of selected mentors. A similar exercise has to be extended to undergraduate students too and in larger numbers. Labs and workshops with some of these subjects need to be set up in polytechnics too, concomitant with new courses. Some of these facilities could be used on sharing basis between institutes. Polytechnics have a greater probability of a culture of building products rather than heavier stress on concepts and theory. Mutual exposure of undergraduate students from science colleges and polytechnics can be rewarding to all, by pervading DIY culture among science students and making them ready for real life situations and better suited to employment.

Academic Incubation and Mentoring

To attain the objective of STEM HE on mass scale, it is imperative that a much larger number of students are attracted to STEM HE and R&D and motivated at an early stage. This requires academic incubation at least from standard X onwards. An associated requirement will be profile monitoring of these potential candidates with the help of their teachers and mentors.

Teachers from mass institutions too need mentoring of a different kind to prepare them for scientific research. For this purpose, each of the national labs and institutes of higher education with high ranking should be tasked to train a selected lot of teachers with specific targets/R&D projects. There needs to be autonomous agencies to monitor and audit their performance. There has been a recent initiative by the HRD ministry to rope in selected IIT students for a similar purpose, but this exercise needs to be made more broad-based covering a larger number of students and teachers to be mentored.

CSR and Increased Scholarships at All Levels

With such a large number of students, there is always a need for much larger number of scholarships and easy loans so that enhanced fee is compensated for. Enhanced fee is required for the sake of improving infrastructure. Government talent scholarships need to be increased in large numbers for all levels of STEM students. But government efforts may not be sufficient and thus industries would have to be persuaded (through industry associations) to offer scholarships to talented and needy students. It could be under the auspices of CSR or otherwise. There are a few industrial houses involved in such missions, but it needs to be encouraged on a much bigger scale. These industries may be more interested in the exercise, if they feel their money is not going down the drain and students are being rightly trained. Some of such trained students could be assimilated by these contributing industries themselves after completion of their studies/research. These scholarships also need to be awarded on a competitive basis. Even senior level researchers working in right directions could be awarded. Few corporates like Infosys Science Foundation, WIPRO and Tata group are setting good examples but their contribution is too meager for a country as vast as India.

Student Projects to Encourage Enterprise

In both science and engineering colleges, normally, some project work is encouraged in the final year of their course. Some IITs have recently permitted their students to join hands with startups, in lieu of their mandatory project work, to expose them to the idea of techno-entrepreneurship and incubators. This needs to be encouraged and facilitated in many more institutes. Incubators and plug & play facilities in STEM departments could further facilitate such goals.

Foreign University STEM Departments

Foreign universities are presently not allowed to bring their faculty as well as innovative teaching/training methods. STEM institutions need to be given special considerations in context of encouraging foreign universities. This may be of good help in several ways in promoting the Indian ecosystem. Their courses (including the open online ones), curricula, number of seats, fee structure and linkage with local industries and other universities could be negotiated in consultation with the best existing institutes. With right choices, it could stimulate competition. Fears of those internal institutions have to be suitably taken care of but not at the cost of encouragement of competition. This may encourage movement towards attaining international benchmarks.

Innovation-cum-Manufacturing Zones

The existing schemes to promote manufacturing in India are oriented to facilitate commercial and logistical issues for industries within these zones. Some of the new manufacturing zones being set up in India could be fashioned so as to encourage innovation and STEM R&D to contribute to manufacturing industries. Such zones should be evenly distributed in different_geographical regions and must be based on either some specialised gap areas or certain futuristic technologies. Each one of them could be centred around some industries/group of industries and close to some IIT (or any other HEI) or its chosen departments having an overlap with these industries. These clusters should be run on international lines, covering best practices. Hsing Hu Science city in Taiwan and some of the Korean Innopolis are good example cases. Funding may be shared between central, state governments and private players in a ratio of 50, 40 and 10 or some suitable variation. Land and infrastructure should be the responsibility of state governments. This policy could coexist with present thinking in NITI Ayog for coastal zone manufacturing for the sake of exports. While coastal zones could focus on manufacturing for exports, the above mentioned zones can focus more on innovation and hi-tech manufacturing.

Hi-Tech Manufacturing and Trade Policies for Made in India

The industrialisation of a country in its true sense can come only from its own resources, both human and natural. This is particularly true for high technology. High technology is intimately linked with latest developments in science and technology. World Intellectual Property Organization (WIPO) defines hi-tech trade to be covering items related to computers, communication, aeronautics, automation equipment etc. There are some more categories which can qualify as hi-tech technologies. Many of the strategic industries related to nuclear, space and defence technologies are also treated as hi-tech by many. Very few countries are self-sufficient in all these technologies. Developing countries will increase their manufacturing, but developed economies will continue to produce hi-tech products with greater efficiency. Thus, for the sake of perpetuation of their own wellbeing, developed countries may not like to part with advanced or high technology. There could be several excuses to cloak this fact. Developed countries have joined hand and created a technology apartheid regime in the name of dual use items and other such pretexts. Their real concern is not always nuclear non-proliferation or weapons of mass destruction, but loss of commercial advantage. This has been proven beyond doubt in the case of cryogenic engine denial to ISRO, where more than missile related concerns, there was fear of a more economic and competitive producer of hi-tech products emerging. Even the Indo-US nuclear deal, which the Indian leadership thought may break this regime, has not succeeded in doing so. Even if India joins NSG, it may alter the situation only marginally. There are so many restrictive lists in operation and trade in them is monitored very seriously. India has been able to successfully beat some of them. Not only technologies, products and instruments are denied, even international collaborations in certain sectors are denied. This way, researchers from developing countries are at disadvantage, particularly in hi-tech research. The Indian defence sector has been made heavily dependent on imports, for reasons of short sighted policies of its own which encourages imports over domestic R&D and production. An often-proposed instrument of technology transfer, FDI, also has not worked for India. India abolished Foreign Investment Promotion Board and raised the FDI ceiling to 49% with the hope of attracting foreign defence industries investing to manufacture in India, but it did not work and India has failed to attract any worthwhile investment. Foreign suppliers clearly want greater control over technology transfer.

How can Indian policymakers incentivise manufacturing based on indigenous research and development in private sector? The first and foremost step is to boost private sector R&D itself and then to offer local manufacturers the requisite ecosystem. In principle, the government cannot enter into R&D as well as manufacturing in every sector. It has to stimulate private sector by various means. There are three major means by which the government can do this. First is by tax incentives, second is by direct support of loans or grants to R&D projects of private sector and thirdly by offering preferential market access. Any policy must evoke confidence in the minds of those who innovate or invest. This calls for stability and consistency of underlying guidelines. Also, it has to be dynamic in response to the fast changing technology scenario. Such policies also cannot be universal and may vary from sector to sector. As technology advances responses of the system has to be agile and even new sub sectors emerge which may call for altered policy response. This may not always be possible as bureaucrats tend to generalise for the sake of simplification of their own understanding of scenarios.

Policy support for high technologies and manufacturers calls for special attention to young startups. Over long periods of existence, incumbents develop vested interests in continuity and thus conservatism. They fear disruptions and tend to influence government decisions in this regard. An example in the Indian context is when the government had to frame a policy regarding product patents for pharma sector as per TRIPS guidelines in 2005, initially there was resistance, but it was followed by a sudden increase in R&D expenditure of major domestic pharma companies (Figure 9 in chapter 4). This was in anticipation of probable competition from MNCs. Evidently, competition is the driving force for innovation. It must be continued endeavor of the government to promote fair competition between various players, public or private, domestic or multinational, big or small. An OECD study in several developed countries established that fair competition is more important than even IP rights or taxation incentives for promoting innovation. The only exception to this rule can be young startups in the budding stage.

Three types of incentive mechanisms have to be suitably calibrated, depending upon the stage of technology evolution in a sector as well as domestic context. Different countries balance differently between tax incentives and direct support. As mentioned earlier, wide variations across countries indicate differences in socio-economic conditions and innovation ecosystem. India predominantly banks on taxation incentives as a support measure. Most incumbents prefer this route while fresh innovators may prefer direct support. Big companies and MNCs manage these incentives to their advantage, unlike young startups. The DST white paper describes all types of incentives offered by the government in great detail [32]. Indian socio-economic conditions and environment, which is not totally corruption free, suggests that taxation-based incentives are likely to be misused and may not be the best course possible. The Government of India has taken a major initiative of mission 'Startup India' to stimulate startup activity in the country. Taking a cue from this, 18 state governments have also declared a policy of supporting startups by various means. The government has also generated a virtual network of linking innovators, investors and government agencies for quick resolution of queries and general interactions. These steps are besides support by way of funding as described in the earlier chapter. This is a good beginning in the right direction.

The third incentive is that related to preferential market access. This implies domestically manufactured products get preference over imported products in government purchases. In case of high technology products, it is quite often that no product is worthy of rigorous qualification as manufactured in India. This situation gets further complicated when MNCs manufacture the same product in India. It becomes difficult to rule out their claims about the product being made in India. Even though they import completely knocked down (CKDs) and semi knocked down (SKDs) kits for final manufacturing. The government has to devise strategies to overcome this to favour young innovative Indian startups. For each sector, technologists specialising in that sector are needed to identify products manufactured by young startups defining the percentage of indigenisation. More important than the percentage of indigenisation is how critical components are part of such indigenisation. Each component has to be categorised as critical, sub-critical and routine with suitable weightage. Upstream value additions like concept, design, components etc. have to be given greater weightage. The government also has to find a way out of the lowest bidder syndrome together with fears of vigilance actions later.

Governments occasionally resort to some kind of protectionism to support their nascent industries. They impose custom duty on certain chosen category of imported products which are being produced by these industries. This may not always be helpful as it is an anti-competition practice. Additionally, those imposing these tariffs and those developing these technologies/products may not be acting in unison. Thus, it does not serve the cause of competitive indigenous industry. An associated issue of significance is that technology does not come off the shelf and there has to be a fair amount of lag time. Thus, the government needs to publicise its hi-tech needs (e.g., acceptance of necessity by the armed forces) in advance to propel industries to compete for orders by pursuing relevant R&D. The Ministry of Defense has started thinking on these lines which is a small step in right direction. This line of thinking needs to be adapted much more widely by different departments as well as state governments.

India needs many of these high technologies both for reasons which are strategic as well as commercial. India as a member country of WTO needs to respect its obligations. As part of TRIPS, it agreed to full compliance with pharma product patents by 2005, even though there was a strong view within that this could harm indigenous industries. There are two aspects — those related to immediate benefits for Indian pharma industries as becoming part of global supply chain and the other is influencing long term competence building. Indian pharma industries are not self-sufficient at all stages of drug development and MNCs are trying to press their gains further by asking for recognition of incremental patents and obliquely evergreening them. While on the other hand, the Chinese, who are lacking in some aspects of drug making cycle, are trying to recruit Indian specialists by attracting them with huge pay packages, with the aim of acquiring industries. It must be remembered that Indian pharma companies source their raw material from Chinese companies. Thus, it is a kind of hi-tech trade "war" being fought on several fronts. India must take cue from Chinese companies that are also trying to acquire hi-tech companies in western nations to overcome the domestic deficit. Two significant examples of this are their recent proxy acquisition of two German companies, Aixtron and Kuka. Such filling of voids in hi-tech manufacturing at national level in each sector calls for a coherent national level strategy involving both public and private sectors. The high technology trade warfare between nations has to be realised as a serious challenge for national development goals. Such warfare between China and USA holds many lessons for several developing nations. Few years ago, US House Intelligence Committee launched investigations into how Chinese telecom companies threaten American interests. Following is a quote from the report [41],

> "We already know the Chinese are aggressively hacking into our nation's networks, threatening our critical infrastructure and stealing secrets worth millions of dollars in intellectual property from American companies. This jeopardizes our national security and hurts U.S. competitiveness in the world market, costing our country countless jobs. The same way hacking can be a threat; vulnerabilities can derive from compromised hardware on which our telecommunications industry relies. The purpose of this investigation is to determine to what extent Chinese communications companies are exploiting the global supply chain and how we can mitigate this threat to our national and economic security."

There was another case years later when six Chinese professors were indicted by the US Federal prosecutors. The professors, who attended the University of Southern California, allegedly obtained the trade secret information designed in part to limit interference in mobile phone reception and other devices as part of a "long-running effort" to benefit universities and companies controlled by the Chinese government. The so-called FBAR technology is primarily used in mobile devices, filtering incoming and outgoing wireless signals. In addition to the consumer uses, film bulk acoustic resonator (FBAR) technology has numerous applications for military and defence communications. The Chinese government has strongly protested against the action. On the other hand, they have complaints against USA of snooping on their own entities. The Chinese government has removed several companies' products from its list of suppliers. These include companies like CISCO, Apple, Citrix, Intel and McAfee. This is being treated as an aftermath of the Edward Snowden affair. The Chinese have charged Americans of spending millions of dollars to snoop on their telecom companies like Huawei. It is reported that as a result of this, the Chinese have accelerated their process of what is called sinification of servers (fuwuqi guochanhu in Chinese). Chinese company Inspur, in collaboration with government enterprises, has succeeded in developing a high-end server named Tiansui K 1 [42]. They have, in a programme called I2I (IBM to Inspur), attempted replacing many IBM servers being used in Chinese banks. Americans fear that after replacing them in Chinese market and consolidating there, China may attempt to sell it in the international market, which will be a huge loss to American companies like IBM and Hewlett Packard. It is believed by many that the present day tariff warfare between the two countries has overtones associated with technology.

Another major case in which the Chinese have been considered guilty of discriminatory trade practices in high technology trade is related to export of rare earths. The WTO ruled that Chinese rare earth export restrictions were discriminatory and intended to achieve national industrial policy goals through controlling the international market in rare earth elements and securing preferential use of those materials by Chinese manufacturers. Chinese supply 97% of all rare earths in the world. India is the second biggest potential supplier, but has not yet made good use of this opportunity, neither in terms of exploration, processing or offering international collaborations. This is related to linkage between rare earth deposits in India with that of thorium a prospective nuclear material. Various kinds of restrictions related to association with nuclear materials have hindered India's efforts.

Some of these affairs related to hi-tech trade do point out to ways in which countries will operate in the coming future. There has to be national alertness, in the form of an integrated response system (particularly response time), to having the right policies and also responding to policies and actions of other governments. Policymakers have to respond and take actions in concert with indigenous researchers and developers and spur specific technology development actions. They have to monitor the movement/immigration of high net patent worth individuals and incorporation of hi-tech companies both at home and abroad.

It is normally pointed out that a strong patent regime can be harbinger of development of new technologies. Most developing nations are not alert to the possible gains or losses accruing to them because of being diffident to progress or lack of progress in patents happening inequitably in different countries. As mentioned earlier, about 90% of all global patents are held by few nations/entities. Governments and innovators in these developing countries have to be sensitised to these issues. The first step is to raise the awareness level of all the players involved by pointing out how the nation will be at loss if it does not generate its own IPRs and protect them. Also, even though patents are generators of new knowledge, they can be biggest impediment to diffusion of knowledge. The healthcare sector and green technologies having bearing on climate changes are the areas which need fast diffusion and deployment of new and more efficient means. While negotiating with international players, developing countries need to be on guard in this respect. Also, within a country, several ministries and organisations have to work together in unison.

National Open Source Register of Indigenous High Technologies

Patents as such are welcome as they are considered engines of growth by incentivising innovators. But the Indian S&T and manufacturing scenario is such that there is plenty of open source knowledge both of local as well as global nature which has not yet been tapped. Steps must be taken to commercialise that first. For this purpose, just as there is a patent register, there is need to have an open source register of high technologies available within the country. They could be collated from its national labs, universities/HEIs, PSUs or private sector. It should be possible for any Indian citizen or entity to easily capitalise on it. This should be digitised and made available on web, with details of how to proceed. Such a step will make present mechanisms more broad-based as preparation for likely future trends. The government as well as innovator community will be aware of gap areas, possibility of synergy and collaborations as well as defining benchmarks against international ones.

The new patent policy announced in 2016 is more aimed at indicating better regulation and compliance for the sake of improving confidence of international investors. It also talks about sensitisation of innovation communities to the benefits of generating and protecting proprietary rights. The Indian patent granting machinery needs a total revamp with a much larger base of patent examiners from among the S&T community to speed up handling of patent requests. To evoke confidence in innovators and investors, specialised and speedy dispute resolution mechanisms need to be established. In India, this will have a major positive impact in attracting talent and capital.

Opening New Sectors to Developing Proprietary Rights

A large part of the Indian knowledge base can be categorised as traditional (like those related to culture, craft, raw agricultural products, traditional medicine systems etc.), informal and government sector research. A major fraction of government funding of R&D is in strategic sectors like atomic energy and defence. These sectors are discouraged in general from holding patents and thus have not acquired a strong culture of patents. These sectors hold plenty of knowledge base and need to be liberated from this constraint and restrictions should be carefully placed only where direct weapon research is involved. This calls for changes in the Indian Patents Act. The new patents policy does talk of awareness programmes about the value of patents. The strategic R&D sector, which is also the biggest employer of Indian hi-tech workforce, will in turn induce many more to be sensitised to value of proprietary knowledge base. As a follow-up measure, spinning off startups based on such proprietary rights need to be encouraged.

High Patent Bar and Revamping Domestic Patent Examination

Developing countries like India must set their patent bar high and not allow incremental research from MNCs to be patented. This should be particularly true for sectors which impinge on their large poor population, healthcare being one of them. This is particularly important as many foreign entities are applying for Indian patents to get an early foothold in a large market. They cover a large range of sectors. Indian innovators and authorities both have to be alert to the situation.

Synchronising Tariffs and Technology Development

Development of high technology involves innovation-based startups and nascent industries. Import duties on hi-tech products need to be handled carefully so that indigenous enterprises get the opportunity to blossom by not facing unfair competition from foreign players. At the same time, at some stage they need to be exposed to competition. This calls for active coordination between those involved in finance policy making and indigenous technology developers. Some of these schemes of incentives fare poorly because of lack of synchronisation of moves. A similar fate may happen to export promotion schemes which may face heat in WTO complaint cells. The proposed NIS could act as an unbiased interlocutor between departments related to finance and commerce on one hand and scientific organisations and technology developers (and industries) on the other.

There are a large number of entities that are stakeholders in several policies that impinge on growth of science, technology and industrial manufacturing. A significant aspect is that unless there is coherence in reforms on all the fronts described above, they will not be effective. Also, while reformative actions on these fronts are highly desirable, these are still not sufficient in themselves. There are major cultural factors that can act as impediments to the ultimate goals and these are not normally given as much weightage as they deserve. The next chapter aims to discuss a few such factors and how poor attention to them may foil all policies and strategies.

Chapter 7

INNOVATION, SCIENCE AND TECHNOLOGY: INDIA'S CULTURAL IMPACT

INTRODUCTION

The previous chapters listed various reformatory action points involving STEM higher education, scientific research, technology development and manufacturing. India is a vast and heterogeneous country with a large number of entities pertinent to these affairs. Government policies, notifications and circulars have limitations in attaining the goals set here as there are several deep-rooted cultural barriers which might need to be addressed. These could be related to traditions, attitudes and mindsets of people at large, society, scientists, industries, bureaucrats, political leadership and entrepreneurs. These set of attitudes cannot easily be influenced by policies and may need actions of a different kind.

SCIENTISTS AND TECHNOLOGISTS: THE BACKBONE

> Most people think it is intellect which makes a great scientist. They are wrong: it is character.
>
> Albert Einstein

How are scientists and technologists as human beings? How are their professions looked upon by Indian society — both by themselves as well as others? It was India's first Prime Minister, Pandit Jawaharlal Nehru who proclaimed scientists to be modern day *rishis* (a term reserved for wise sages in Sanskrit). Does it continue to be so?

Science has been defined in several ways which are essentially several facets of the subject and those who practise it. One famous remark about science, distinguishing it from other professions, was made by Richard Feynman, the famous Nobel physicist. He mentioned that while unquestioned "expert opinion" rules the other fields, in science, the experts of the past are ruthlessly questioned and frequently toppled. In history or economics, these things happen on a much larger time scale. Many of the working scientists must have experienced this in their lifetime, while others were refuted after they were no more. Lesser mortals face much worse. But science is not a slave to the ego of high and mighty. Feynman further characterised science to be a belief in ignorance of experts [43]. He distinguished between what he called real science and fake science: real science always involves extreme bend-over- backwards honesty. Fake scientists get defensive when criticised and try to hide their mistakes, while genuine scientists attack their own ideas in order to make sure they are correct. Also, real scientists encourage critics to attack those ideas and help their critics. Their aim is to distill the real truth and it is immaterial who wins the debate. The question today in Indian science is what kind of science is being encouraged? There are many other attitude and behavioral aspects which are commonly expected from people of the world of science. Curiosity and questioning are common to most workers in these areas. Passion and tenacity to pursue an idea, like in the case of Madam Curie, are rare but are ideally expected from scientists. Systematic observations and data handling, like that of George Abbot Mendel — the father of discipline of Genetics, and many others of the same ilk, are exemplary. The courage of Galileo in standing for the truth even at the cost of personal welfare is rare, but has inspired generations of scientists. How many scientists in India follow these ideals today?

Most of the scientists are taught only the subject of science, engineering and technology. The above mentioned level of passion for truth is not taught anywhere, but is imbibed from the milieu around, if it exists. There are very few who are born with it. This is a process which is defined by the leaders in renowned science institutes. It comes by example in behaviour rather than by preaching. It is wrong to think these aspects are not significant in contemporary times. Many in Indian science today think that pragmatism is a better virtue. Frequent cases of plagiarism in Indian science are noteworthy in this context. Many of them involve the so-called leaders. There are so many ills which show up in how science is practised in India and the standards followed. It is not realistic to expect that all these qualities will be found in all scientists. But absence of some of these features, at least at leadership level, can affect healthy growth of science.

Silence of Lambs

Scientists in most government labs are also government servants and not normally professors. This places on them responsibilities and constraints related to rules of discipline ordained by the government. This much is understandable. But quite often it transforms into silence of lambs, which is more related to their personal welfare.

It was Homi J Bhabha, the great visionary, who while pursuing the idea of an institute of fundamental research, advocated that "if much of the applied research done in India today is disappointing or of very inferior quality, it is entirely due to the absence of sufficient number of outstanding pure research workers who would set the standard of good research and act on the directing boards in an advisory capacity." This statement was made a few decades ago, but is even more relevant today. He expected these pure researchers to have a vision for future and have the courage to speak the truth in matters related to various projects and proposals of science and technology. India needs more such scientists who can speak the truth with courage. Many Indian labs today have put "discipline" on premium and individual scientists have put their career on premium. Many of the heads of these labs tend to fill such advisory committees with scientists who are of convenience to them. All this is at the cost of genuine growth of science and technology.

Many such qualities do not figure while evaluating them. It may be worthwhile to have their performance appraisal in such a manner that there is a mention in brief about how in practice they adhere to the 'spirit of science.' In postgraduate STEM education, there could be courses on scientific ethics and history of science and technology to inculcate the requisite value system. This will upgrade the quality of human resource in general and has potential to impact significantly the performance outcomes. One must not forget that scientists and technologists have potential to be the true backbone in building a nation.

SCIENCE, TECHNOLOGY AND BUREAUCRACY

> A scientist will never show kindness for a theory which he did not start himself.
>
> Mark Twain

It is very often that the Indian scientific community blames 'bureaucracy' for lack of performance on their part. It is worthwhile to examine how this so-called bureaucracy arises and how this comes into play. Scientists and technologists have to interact with non-technical personnel to facilitate and sustain day-to-day running of their labs and institutes. They have to obtain funds for their projects. Depending on the magnitude of the project, the approving authority is higher up the ladder of administration. They have to obtain land and other infrastructure requirements, for which again they have to depend on the administrative (or political) authority. They have to get their staff (manpower) vacancies sanctioned, promotions and high-level appointments from authorities who are not always scientific. Scientists have a mode of presenting, asserting and propagating their results — by way of publications in journals, conferences and patents. They do need support for this from time to time.

Additionally, many a time, research is of multidisciplinary nature and may require collaboration with some other national or international lab. This may require sanction or support of a different kind from various departments/ministries involved. In most of these activities, scientists have to interact and depend upon bureaucrats as well as political authorities.

Bureaucracy or red tape is something which all scientists fear for its capacity to cause deliberate hurdles in R&D. Bureaucracy is normally supposed to provide some check on unbridled behaviour and guard against misuse of funds without themselves having even a little understanding of a scientific project! They are expected to implement policies consistent with best practices. It must be questioned whether professional bureaucrats have succeeded in these goals or not. Poor handling not only affects the task at hand but also the morale of researchers and creates bad vibes. The 'idea' of a researcher/developer/innovator has to traverse a long journey before its fructification to be of useful result/product. Mostly researches in large government scientific institutes in the country are sponsored/supported at the level of organisation itself. Either the innovation idea originates in the mind of a young researcher, a doctorate, a post doctorate or a professor level worker. This idea could be related to the mandate of the organisation/institute or social mandate and might have overlap with the ongoing research area of the staff. The other way for the idea to take shape is that research manager/mentor at the top dictates/convinces the researcher to carry forward research in a particular area of interest to the organisation. The journey of an "idea" is quite different in two cases.

In the first stage, the plan is put forward as a proposal to local organisation. It goes through some scrutiny, in various fora within, by different scientific committees. If the idea is sponsored by the management, it goes through without many hindrances. But if it is an individual level proposal, it has to overcome several barriers. It has to be defended at higher and higher level depending on how much funding and other support it requires. Mostly these committees are filled with scientific experts and occasionally have a representative of finance who rarely has the power to overrule the scientist chairman. If the funding/support required is still higher, then the organisation has to get approval from, say, ministry (earlier Planning Commission) or secretary of the department to check its consistency with overall policy guidelines and facilitate related matters. It is true that large sized purchases are delayed by way of procedure consistent with finance audit. Even secretaries of many scientific departments in India are mostly from science cadre and are quite powerful (administrative power wise).

In the case of a researcher at a university or other institute of higher learning, both public and private, he/she has to get initial approval from his supervisor to be followed by head of department and then has to seek funds from various funding agencies in the country, provided the university is willing to give him/her support in terms of infrastructure and manpower required. Many a time, even the local department head becomes the hurdle to project implementation by denying the principal investigator basic amenities. These funding agencies have their own 'strict' (many a time not transparent, as it seems to proposer) yardsticks for sanctioning of funds. Since private sector is not yet prepared to enter the arena of research on a large scale and Indian university researchers are not involved in 'semi commercial' (or industry-related) research, these researchers have to fall back on government funding sources. Mostly the committees expected to review these proposals are constituted of eminent scientists from within the country. In fact, experience with government funding agencies for universities shows that on most occasions it is a case of sufficient funds, but insufficient researchers, particularly in areas which are applied in nature or are challenging. For example, in case of DST, the Science and Engineering Research Board is supposed to promote R&D in various disciplines of science and engineering by giving funds to researchers from across the country. To facilitate their work, this board has constituted various programme advisory committees with specialists from all over the country. It has sanctioning powers to a certain extent and for proposals involving larger funding, they are forwarded to the board with comments or recommendations of their own. Even at the time of closure of the project, the researcher has to come back to the same committee with a report about the research results. All these sanctions do not involve any political person, administrator, accountant or auditor. The process is same for many other scientific departments. Some are more autonomous than others like AEC and space commission that are answerable directly to the Prime Minister. The response time of approving and sanctioning, fairness and transparency of the process, is related predominantly to scientists and technologists and not professional bureaucrats. There is always scope for procedural simplification, which can be decided by scientific personnel. Recently, the Government of India has announced the formation of an inter-ministry group which will sanction about a billion dollars for carefully selected project proposals from various IITs and NIITs. This is expected to resolve some of the complaints about bureaucracy, but implementation will be the key. A large number of project proposals were prepared, but a few months later, many of the departments and ministries have developed cold feet in sharing funds. This type of dragging of feet is partly related to poor credibility of the scientific community in general to deliver up to the last mile.

With this level of independence enjoyed by the scientific community, still there are complaints related to the bureaucracy hindering scientific progress. Then, of course, there are bigger questions like policy on total expenditure on R&D as a fraction of GDP and what type of research to be given more weightage. These are questions discussed at maybe Niti Ayog level (high level policymaker group/think tank) which have to be faced by scientific managers. This is the interface of public accountability of Indian science. Unlike earlier Planning Commission, Niti Ayog is more advisory in nature and is not a fund disbursing authority. This body is heavily dominated by bureaucrats and economists.

Anything other than basic research has to have some end user. It could be industry, public (in the case of power), health or agriculture-related R&D, defence sector or the armed forces for indigenous development of weaponry, satellites and launch vehicles for communication or remote sensing, navigation etc. All these end users have a right to question the scientific community responsible for developing these technologies. The yardstick of success has to be defined by several stakeholders vis-a-vis what happens internationally.

Nowadays, in India, there is talk of a catchword, 'innovation,' in new contemporary context. It is often demanded that the bold new world of research with radical ideas should not be audited. In R&D, failures are more likely and too many questions are looked upon as 'bureaucratic oppression' by scientists. To the public mind, the question is how to differentiate between 'genuine failure' and 'down the drain' research carried by some incompetent researchers. If there are professional regulatory mechanisms in place, then 'outsider bureaucrats' need not interfere with the process of audit, when the audit is not purely financial but related to physical progress on the ground to the satisfaction of the end user. Also, not only projects are questioned milestone-wise, but even choice of R&D projects can be questioned. Even for the private startups, there are certain self-regulatory mechanisms. The average rate of failures and exits vary in different countries. India needs to strive to attain the global average of exits. Government funded R&D must have some self-examination and outsider monitoring agencies can be healthy for growth of the sector, provided the appraisal is from genuine outsider technical experts, besides the user groups. Any malpractices in such reviews are one of the biggest hurdles in the path of progress.

But there could be genuine researchers/developers/innovators who may still have complaints regarding 'interference' in their work. This 'interference' is also called bureaucracy. As is evident from the above discussions, most of this 'interference' comes from within the scientific community that sits on the jury's seat to decide the fate of all 'new ideas.' Right from the stage of acceptance of new ideas to sanctioning of funds, giving institutional and infrastructure support to propagating research results and final performance appraisal and closure of project, it is a process that goes on within the scientific community. If the scientific community is able to attain all the objectives with full accountability, then these bureaucratic controls may not be justifiable. Not only that, sanctioning is important, but it should be timely and smooth as delays can be fatal for frontline projects related to high technologies. One must continue to ask what actually happens that hinders the progress of new ideas and prevents them from taking the shape of a new product/technology. Will appointment of a high-powered ombudsman to monitor and regulate help?

In an ideal world of young scientific researchers being encouraged by rest of the scientific community, there should not be any grudges or complaints. But it isn't an ideal situation. Scientists and technologists are also human beings with frailties like ambition, jealousy, groupism and so on. Today's laboratory is not a simple one like that of old times, dealing with just test tubes or galvanometers. Today's laboratory requires complex and costly setups and equally matched test instruments or facilities. Scientific research involves large teams of multi-disciplinary manpower. The career growth of scientists and technologists comes to be linked, in some way, to how much funding and manpower one controls. Even the gestation period of these projects-cum-experiments is large. Inadvertently, the controlling scientist/technologist reaches that position not always by his track record, but by his probable potential to deliver. There is thus a rat race to grab funding and power just as in many other professions.

Many a time, science managers responsible for sanctioning funds are either ignorant of complexities of the project they are sanctioning or are gullible (or plain dishonest) enough to be convinced by 'over-smart' scientists proposing that project. This is not the only malady. There could sometimes be genuine perception difference between the proposer and the sanctioning/approving authority. But at other times, these differences could be faked for the sake of some alternate 'non-technical' motive like prejudice, part of a group or lobby fight or even an insecure leadership suppressing a budding star performer. In each of these cases, the institute is at a disadvantage. Thus, while doing performance appraisals, failures of 'favourites' and successes of contrarian 'lobby' are ignored. These 'lobbies' could be sometime even based on subject, with the science manager favouring his own field of work at the cost of some other one which may be deserving more attention for reasons of feasibility or synchronisation with national goals etc. Also, most of the innovations happen as a result of interdisciplinary 'out of box' thinking. With lobbies working for each discipline or field of work, little will be left for new upcoming interdisciplinary and innovative research. Sometimes, the 'fight' is between traditional way of thinking and a radical approach to the problems. In fact, this fight happens all the time in science and technology, even at the global level. This is because inertia and preservation of status quo are very powerful forces in themselves to defeat (or thwart) radical and innovative solutions. With increasing pressure to march with 'innovation driven' R&D, there are demands that failures should be ignored as it happens in all high risk 'innovative' ventures.

In an atmosphere of the kind mentioned above, it may not be prudent to give an absolute free hand to all science managers without some suitable regulatory yet non-bureaucratic mechanisms in place. This kind of atmosphere discourages genuine researchers/developers/innovators and performance of individuals as well as institutes deteriorates fast. A more significant concern is the influence it has on 'atmospherics' for the young researchers. It may be worthwhile to recall that in Korea, most of the science students or researchers who go abroad for higher studies return after a certain period. This has not been so for the case with India and China [40]. But China, in recent concerted drives, has been able to bring back a large number of its students.

This 'atmospherics' is branded as 'bureaucracy in science' by many. Thus, scientists themselves can become a hindrance to the progress of their own kind. As can be seen, the bureaucracy about which many scientists complain is not the same as departments of finance and administration. In summation it must be asked that whatever bureaucracy exists, internal or external, scientific or non-scientific, has it been effective in producing the desired output?

INNOVATION, PAPERS, PATENTS AND PRODUCTS

Innovation manifests itself in different forms. It could be scientific research publications, patents, novel products, processes or even novel organisation management or marketing. Just as input of human capital is difficult to quantify, similarly, research output is not easily evaluated.

While research publications are indexed nowadays by citations, patents are known by their commercial worth or extent of usage.

An OECD study [44] on S&T outlook for India defines few broad parameters like science base, business R&D and enterprise possibilities in the country as the basis of progress in terms of S&T for manufacturing. The first one, the science base, is further categorised by three factors — public expenditure on science, top 500 universities and publications in top journals all normalised to GDP. The business R&D is categorised based on three factors — business R&D expenditure, top 500 corporate R&D investors and triadic patents (patent in US, EU and Japan) and trademarks again all normalised to GDP. The third one, the enterprise possibilities in the country, is related to ease of setting new enterprise, venture capital support and patent-based firms (less than five-year-old). Besides these parameters, development of human resources based on the level of science education in the country and percentage of population in S&T, the knowledge flow which is related to patents filed by universities and public research labs and industry expenditure on public R&D are other significant indicators. They compare performance of India against OECD countries and their median output in each of the parameters.

Figure 13 shows some of the parameters of comparison on progress of S&T and status of India with respect to OECD median [44]. On most of these parameters, India performs poorly, but the parameters where performance is quite bad are human resources like doctorates (science & engineering), percentage of S&E manpower of population and patents and industrial expenditure on public R&D. The current Indian government has worked hard to improve ease of doing business and it has improved India's ranking substantially. It can be seen that this parameter is within the reach of non-scientific policymakers, but most other parameters are dependent on science managers, business leaders and those responsible for STEM HE.

The mission 'Made in India' with an objective of designing, developing and manufacturing in India with the help of Indian trained workforce demands an appropriate mix of these innovation outputs. Poor status of some of these parameters mentioned above, vis-a-vis several developed countries, necessitate actions on several fronts.

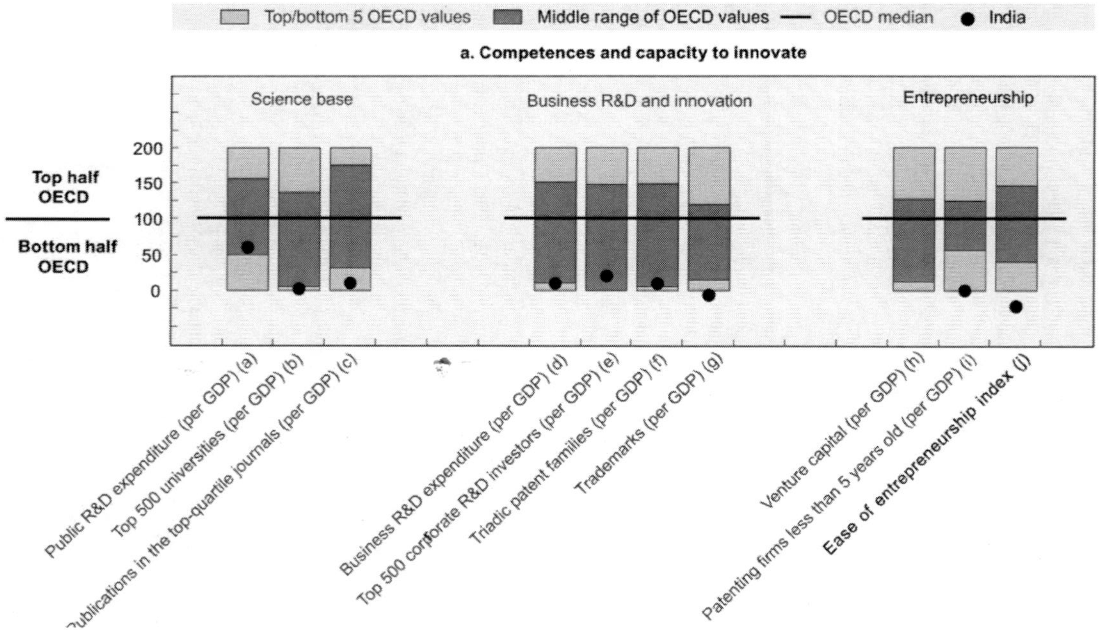

Figure 13. Comparative performance of India and OECD countries in terms of innovation and enterprise. Various related parameters are indicated.

But these are not sufficient in themselves as culture dominates heavily against these structural reforms. Figure 13 describes only statistics of enterprise development based on scientific research and technology development in a country, but does not explain the origin of these problems, for which one may need to dive deeper into the cultural domain [44].

Most published scientific papers are not an end in themselves and do not always end in the hands of an end user within India. Even publications with good citations either end up as what is called 'excellent contributions to science' or are made use of by external entities or industries. Such wayward progress in science may offer satisfaction to individual workers, but it does not contribute to national development. Any growth statistics is meaningless in the absence of a coherent thrust. A large number of Indian patents (by Indian entities) also do not find applications in manufacturing. This is not to deny the small local purposes in scientific research being served well. Culture plays a major role in such phenomenon and such influences are normally underestimated or ignored. A stark example of this is America owning patents on herbal remedies with roots in ancient Indian Ayurveda (ancient Indian system of medicine). Sir J C Bose on wireless communication is another glaring example of attitudes and mindset — missed owning these technologies of wireless communication despite his pioneering contributions. India has been responding and reacting to some specific situations on case by case basis, but is still far from imbibing that spirit of aspiration to master and possess technologies.

It may be wrong to conclude the above discussion as belittling basic science, whose forte is mainly publications and citations. Basic science has a big role in visualising development goals, train human resources appropriately and be responsible for inculcating a culture of innovation. In fact, all technologies have their seed in basic science, which acquires a purpose beyond curiosity at some level of growth. Quite often young Indian researchers develop the misconception that R&D can be 'either paper or product' but real experimental research is all about 'paper and product.' Both of them need to grow hand in hand. The real depth of R&D is reflected only thus. Such attitudes also stem from cultural milieu.

Conception of new ideas, designs, evolving trademarks and brands and staking claims to patents — all these are parts of a ladder which must be scaled by Indian researchers. The challenge for Indian S&T ecosystem is to make scientific research more purposeful and find its true place in national development. Scientific research and technology development in India have to confront twin challenges of evolving hi-tech manufacturing for the sake of higher quality jobs, for its millions, and face the onslaught of futuristic technologies. This requires a major change in attitudes, besides policy changes.

Innovation has to dig deeper in Indian soil. Presently, innovation at ground level in India manifests itself as part of necessity of survival, occasionally reflecting as *jugaad* (contrivances born out of necessity) or frugal engineering. This tends to be localised and does not get wider exposure to market/commerce. Harnessing innovation demands venturing into risky areas. The economic insecurity of a large fraction of population discourages risk taking. This is also reflected in scientists, technologists and innovators in general. Most of the risks taken are of incremental kind. This is true equally for those in public or private sector. Young innovation-based startups are a recent phenomenon, restricted mostly to service sector, has yet to spread to manufacturing and spread widely. Even the scientific workers in government sector mostly build their careers on papers rather than patents. Product building is considered not rewarding enough. A major cultural lacuna is hesitation to puncture silos, collaborate freely and go beyond rigidly defined mandates. Developing products or patents is not regarded highly among the Indian S&T community. In most countries, there is a small variation between big industries and MSMEs about their capacity/capability to innovate. But in India, MSMEs are very poor in terms of innovation capabilities as they employ underpaid and low skill workers. This is significant as they form a major part of manufacturing workforce of India. This behaviour sharply contrasts against some of the technologically developed countries like Korea, Germany, Sweden and Norway.

Patents data in a country is often treated as an indicator of trends about linkages between innovation, scientific research, technology development and manufacturing. There are few noticeable trends in Indian patent scenario which are worthy of mention here. If one examines the data related to patents filed and granted in India, then some observations are noteworthy. The number of applications has been increasing at good pace, but those granted are much less and the number of patents granted to foreigners is at least four times those granted to resident Indian nationals [25]. Most of the patents are from metropolitan cities, with the maximum being from the state of Maharashtra (Mumbai-Pune). One needs to remember that India's real potential reflects through its numbers and spread of patents culture over different geographic regions is imperative for healthy and inclusive growth. Another issue of concern is sectoral contributions to this list of patents. OECD World Industry Service has categorised patent related activities into three categories. Public KI (knowledge intensive) activities are those related to education and health, commercial KI activities are those related to business, finance and communication. The third one is that related to high technology which covers aerospace, communications, semiconductors, computers, pharmaceuticals, test and control equipment and office machinery [45]. IHS Global Insight World Industry Service database indicates, as shown in Figure 14, progress in terms of levels of technology and knowledge base in these areas of some of the important economies [45].

It can be seen that India performs reasonably well in KI industries, but poorly in HT areas, while China retains a good balance in all three sectors. Thus, at the present stage of development of Indian economy and S&T, there needs to be robust growth of high technology patents and product development both region wise and sector wise.

The last but not the least is poor comprehension of nature of linkage between innovators, investors and manufacturers. Venture capital, incubators and accelerators and appropriate mentoring of young entrepreneurs is integral to healthy and innovation-led growth. A large number of young startups are registering themselves outside India. This indicates the need for evolving a culture and consequent actions that values each player.

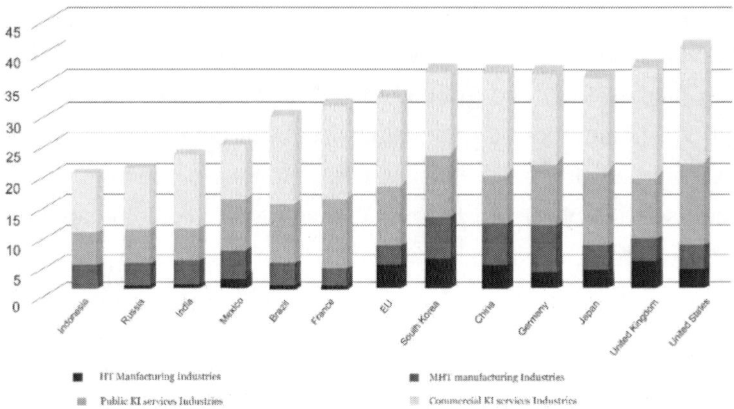

Figure 14. Comparison of major countries, in terms of percentage of contributions to GDP, in levels of knowledge and technology intensive industries.

Developed countries ensure that strategic technologies of yesterday based on patents are released today for market and returns from this are ploughed back to fund strategic technologies of tomorrow. But in developing countries like India, funds today are spent on buying technologies/products of yesterday with little 'thought' and resources (innovation) for tomorrow. This is interspersed with nominal sprinkling of R&D efforts, thus they continue to chase the mirage of becoming developed, at elephantine pace. Unless a cultural movement, that removes bias against products and patents, is brought about, this situation will continue to remain so.

HUMAN RESOURCES FOR DISRUPTIVE TECHNOLOGIES

> We believe that if men have the talent to invent machines that put men out of work, they have the talent to put those men back to work.
>
> J. F. Kennedy

One can only hope this is not a false prophecy. Disruptive technologies have to be viewed in the light of demographic profile in several developing countries. What is demographic dividend after all? Many of the western countries that have been having falling rates of population growth for quite some time have suddenly realised that their population is aging fast and soon they will be short of workforce and older people will be a burden on their infrastructure like healthcare etc. They have come across data which points out that many of the emerging economies have median age of their population which is much lower than that of many developed countries. This is often projected as the demographic dividend of these economies. The moot question is will this dividend really deliver itself against the onslaught of disruptive technologies which aim to replace cheap human labour with machines. India is one of those countries that have a similar advantage of dividend and this may continue for few more years [25]. Thus, India faces the challenge of reconciling these seemingly contradictory trends, viz. development of disruptive technologies and harnessing demographic dividend.

In India, for a long time, population was projected to be a major problem — it still is! This was particularly so in the era of acute shortages of foodgrain and so on. Now, some of the Indian policymakers have woken up to the reality of a possible dividend. The reason for not being able to make use of it is related partly to poor comprehension of the complex nature of the problem and partly as it requires a multi-pronged initiative at a very high level and involves several public and private organisations and departments. The problem of unharnessed population could be aggravated to serious proportions because of oncoming influence of disruptive digital technologies that are unfolding faster than expected. Also, it is to be seen whether India (and many other developing countries) will be able to provide the right mix of skills and higher education to its burgeoning population. If radical steps are not taken urgently, then population is more likely to remain a problem than a dividend. In this context, India and many developing countries must realise that for a variety of reasons, they are racing against time, unless they act fast and right. The key ingredients being STEM HE and the right mix of scientific research and technology development.

Unpredictable Technology Developments

> We can rarely see far enough ahead to know which road leads to damnation. Whoever concerns himself with big technology, either to push it forward or stop it, is gambling with human lives.
>
> Freeman Dyson

There are plenty of new technologies on the horizon which can have a disruptive impact on what one calls the traditional economic scenario. This is particularly alarming as these technologies are likely to have an impact of employment. An example often quoted in the context of US economy is Instagram, a photograph sharing site. When it was sold out to Facebook few years ago, it had only 13 employees and 30 million customers and was sold out at about 1 billion dollars. In contrast, Kodak, a company in the same business segment (of pre-digital era) employed, at one time, more than one lakh employees, had to file bankruptcy papers. What was Instagram technology yesterday could be artificial intelligence, robotics or machine learning tomorrow and so on. This is more a reflection on the march of technology and social behaviour than on the companies themselves. What happens to developing countries like India where predominant populace is still involved in inefficient processes dominated by small industries and agriculture. Their workforce is coming out to join manufacturing and service sectors but at very slow pace as the majority lacks education/skills required by these sectors. IT and IT-enabled services are causing creative destruction of many traditional jobs. AI and robotics may aggravate the problem manifold. Scientists and technologists can probably only predict the type of changes that are likely to happen in the future technology scene. The famed debate about differences on development of AI between Elon Musk and Satya Nadella is an example case. A solid contact with ground reality is required to be able to predict what is in the domain of possibility for developing countries, compared to international developments. There could be strong variation in the impact upon developing and developed countries. It can also point out the potential for Indian workforce vis-à-vis that in many countries as this will be the crux of a probable demographic dividend. The digital revolution caused by many companies like Amazon, Google, Facebook, e-bay and others have created a large number of new jobs. They have offered a huge new set of services to consumers. But most of these jobs require high level skills. At the same time, these companies have killed many existing jobs. There are several emerging Indian equivalents which may not solve the problems related to employment of existing skill level workforce.

It could be worrisome for developing countries the way new technologies like AI, robotics, 3D printing/additive manufacturing, IOT are evolving. Even the developed world is finding it difficult to have the right kind of workforce. It is said that even in Europe and USA, educational institutions are not producing the requisite quality and quantity of manpower which advanced manufacturers need [37]. If this is the situation there, India is going to suffer from much more serious skill deficit nullifying all opportunities of harnessing the demographic dividend. Traditional management of expanding institutes of vocational education, polytechnics and the like, will not suffice. There is need to identify these new-age skill sets and seek novel means of training. This is a gigantic task because of the large number of low skill workers and inadequate human resource development machinery. Many premier R&D institutes that have exposure to some of the relevant advanced technologies need to be coopted for their training potential. This alone may not suffice as the numbers needed to be trained may increase rapidly and a vast variation in educational backgrounds may require some other mechanisms and agencies (both private and public sector) may have to be involved. There is great scope for EdTech startups to enter the arena. Even international Massive Online Open Courses (MOOC) can be involved. More significantly of all is the fact that vast sections of Indian society in general are slow to adopt and assimilate new technologies. There are hesitations at both societal and political level. This arises from the fears of potential job losses. This inhibits the society as a whole to respond appropriately. This was so at the time of the advent of computers too. While in this case, the response time is crucial.

This situation could be both a challenge as well as an opportunity for India. Its large population, wide spectrum of workforce, basic strength in IT sector, a fairly strong R&D base in many relevant areas and a large market can serve India in good stead. There is need to generate a symbiosis of all these segments which is missing presently. India and many other developing countries will do well to sharply focus on generating STEM workforce in the right mix of higher education and skills. Just as there is a spectrum of capability requirements, so is true of variations in the potential of its workforce. Matching the available skills and education to demands is a challenge and it requires strong will at the highest level to remove various kinds of roadblocks and ensure strong coordination between various agencies involved, as mentioned in the previous chapter. Only this can activate the requisite support to education sector to be in link with both R&D sector and advanced manufacturing sector regarding the type of education and the number for each category. This may require weaning away the young from relatively worthless 'degrees' to highly useful 'diplomas.' OECD countries have an excellent mechanism to match supply of skills and their demand based on robust and dynamic data collection. Developing countries like India may have to develop a robust database of its manpower to attain optimum utilisation of its human resources. Indian cultural preferences in employment have to make way for modern demands. A right mix of researchers, technology developers, designers, knowledge workers and new age skills may reduce problems of employment and maximise benefits of technology advancement.

KNOWLEDGE BASED GROWTH: INDIAN RESPONSE

> Technology feeds on itself. Technology makes more technology possible.
>
> Alvin Toefler

What is knowledge based economy and what is its role in present day world [46]? The subject of discussion at present is what is its role in progress of nations, what has S&T to do with this and how developing countries like India are geared up to face it. All the progress of material world is related to progress in S&T. It was industrial revolution in the nineteenth century which was the first harbinger of a new era based on S&T which influenced nations in many ways. It increased demand for new products which were manufactured in countries with the technology to produce the same. But the raw material and commodities required were sourced from a different set of countries which had these natural resources, but not the technology for producing these products. Finished products were sent back to these developing countries with huge profit margins. Thus, technology gave birth to colonialism. With time, many less developed countries acquired (or developed) the capability to produce these same goods. This reduced to some extent the differentials in wages and living standards of these two categories of nations. Present day developments related to digital technologies and the likely advent of industrial revolution 4.0 may tremendously enhance the value of knowledge over manual labour. New type of stresses may emerge across several developing countries.

New Age Colonialism

Developed and developing countries are in some kind of an unequal race to master new knowledge base and technologies. This might have looked purely like an unequal race, but the emergence of soft skills and their impact on various aspects, in principle, has the potential to offer some hope to developing countries. This time the currency is not simply manufacturing and hardware but 'know how' or knowledge. Much of this is related to softer skill set, many a time linked to manufacture of advanced technology products. Software codes, designs, trademarks etc., also labelled as KBC, are playing an equally important role in competitive manufacturing of new technology products. For example, autonomous vehicles need computer-based skills more than simple automobile engineering. If this is coupled with electric vehicles which require even fewer components, the software component of futuristic autonomous electric vehicle may be larger than its hardware. Knowledge-based innovators are likely to have the upper hand. This gives an opportunity to developing countries like India with prowess in software to play a much bigger role.

It must be remembered that for the developed world, it may be a struggle to maintain their level of living standards, but for countries like India, it is a struggle for survival. This is because with knowledge as the basis of development in future, this may increase demand for high skill labour and marginalisation of low skill labour. Keeping in mind its large populace, which is unskilled and may soon prove unfit for 'high skill and knowledge-based growth,' it may cause major disruption in the process of inclusive development. Developed nations on the other hand have a different perspective on this and they always consider the competition they are likely to face from low wage but medium to high technology nations like India and China. They also have the power to attract talent with knowledge base from these nations, bringing developing countries again in trouble.

How the science system comes into picture in all this? Knowledge-based growth is heavily dependent on few basic aspects — new means of production and transmission of knowledge of S&T as well as training of related manpower. Each of these is intimately linked to scientists and technologists. Novel means of production-based on knowledge are all related to R&D, but the new demand on these knowledge producers is to link up with demand generators — the consumers, market and industry. This is taking scientists away from their favourite subject of 'know why' or curiosity. For Indian scientists and technologists to get into a new frame of mind requires some change in attitudes and persuasion by way of appropriate policy framework. This new mindset demands from science and technology workers to not only be in touch with the frontiers of science and technology, but also with the pulse of market. This may also necessitate non-S&T workforce, like management professionals, to supplement efforts of scientists and technologists in fine tuning the outcome of R&D to market requirements. Policies related to S&T call for weightage or stress on different layers of science (like basic, applied or technology) and these have to be readjusted with sector and time. In fact, at different stages of growth, each of these frontiers will become crucial. The effort of translating ideas to lab products to market place is crucial and often neglected in India. Different countries have different levels of capability on this front. This is particularly true for countries like India that have a fair share of global R&D (output), but not in hi-tech manufacturing. India needs to learn the skill of 'lab to market place' to enable good returns on its investment on R&D.

Innovation is enabled by technical competence and advent of new technologies existing together. Developed countries have at present the strength of 'capital,' but the limitation of 'innovation at high wage' and they fear losing their hold to countries having knowledge workers at low wages. They will fight aggressively to retain their hold on technologies based on existing patents which may last fairly long. It must be remembered that at present, collectively they hold 90% of global patents. They will all like to have this strength to be evergreen. Since they are capital rich, they can buy a lot of R&D (or employ knowledge workers) as a commodity. They can buy out big companies with knowledge-based capital too. Countries like India must awaken to their potential of having low wage R&D rather than only cheap manual labour. This should lead corporates from the developed world to invest on R&D in India. As mentioned earlier, a few thousand R&D centres have been opened by MNCs in India with billions of dollars pumped in. India at present is considered to be the most attractive destination for MNC R&D centres. This needs to be leveraged suitably to local advantage. At present they do lower end of R&D and make use of the local tax structure and do not share their R&D results with local entities freely and use this R&D for their manufacturing back home. It must be remembered that the entity with greater hold over knowledge and view from the top is in a better position to turn it to its advantage. As the new wave of 'next shoring' arrives, MNCs may seek pastures where the demand is and where the talent is. This again points out to the necessity of attracting and retaining talent in developing countries against competition from developed countries. Hopefully these foreign R&D centres will exploit so far unharnessed talent better (or at least supplement) and develop distinctive type of products demanded by the domestic market. At least by offering competitive wages, they will be able to curb (or reduce) the lure of foreign lands for better prospects for S&T researchers. By bringing such R&D-based products in the local market, they may establish the worth of R&D in the minds of local industries and public. If handled properly, this can be a win-win situation for all. Most importantly, this may also encourage the spread of culture of taking up R&D as a career path.

Many of Indian pharma companies have done wonderfully well. They seem to be posing some sort of threat to the established international order in the pharma sector. Some of it has usefully employed principles of 'reverse engineering' and cheap workforce. This needs to be replicated in many more sectors. Reverse engineering need not be looked down upon as is many a time done by some. In high-tech sector, this could be an extremely complex and intelligent process. Reverse engineering and frugal engineering will play a significant role in Indian manufacturing sector and several aspects of knowledge-based growth.

It is wrong to presume that knowledge-based economy (KBE) means advent of ‘soft skills’ at the expense of ‘hard skills.’ But it may need a combination of the two. Advanced technological products will be increasingly dictated by softer skills and manufacturing may acquire an altogether new complexion, 3D printing and internet of things are some examples. Human capital may become more significant than the conventional ‘capital’; also, human capital has the potential to appreciate with time. There will be need for lifelong learning for majority of workforce as there will be fast obsolescence of technologies. Those who do not understand the complex nuances are likely to falter and fail. This calls for active involvement of dedicated technocrats, covering various disciplines, into policy making structures. These individuals can monitor the situation closely and respond appropriately. They should not be simply in advisory positions (when they tend to be yes man), but in spearhead positions. The response time of the system has to be fast and responsive to fast changing technology scenario. This will demand major cultural adjustments from Indian job seekers too. Employers also may have to adjust to make provisions for in-house training/orientation. The government may need to support employers in these reorientation efforts.

Along with developing technology, India needs to project a humane face of the same. Any modern technology or manufacturing, in principle, does some violence to the environment. Additionally, there is still a large part of the population which is impoverished, under nourished and exposed to extremes including deteriorating environment. This calls for ingenuity on part of Indian innovators and researchers. What technologies will be sustainable in the long run with such constraints? Developing countries will do better to uphold the Gandhian principle of "small is beautiful," which is also espoused by famed economist E. F. Schumacher (though all related ideas may not necessarily be followed). This is more likely to be consistent with the Indian domestic situation as well as principles of sustainability, efficient use of resources and inclusive in nature. They need to face the challenge of their own model of development in the long run. For example, the technology of 3 D printing or additive manufacturing uses raw material more frugally and offers dispersed mode of production and economic repair of costly equipment instead of necessity of 'use and throw.' Repair of complex and costly equipment rather than replacing it – again suits lifestyle and economics of developing nations. This technique has also been suggested for applications like low cost and faster building of civil structures like housing, bridges etc. [21]. Similarly, India has an ambitious plan of linking all its villages by the internet — a mission with acronym BHARATNET. This can be very usefully deployed where higher level knowledge workers can remotely monitor and control operations by relatively lower level knowledge workers, say in inaccessible locations. This could also be 3D printing of various types of products. It aims to empower village communities in various ways. It tends to make village communities less dependent on supply chains. Thus, the need to have a large number of widely distributed machines of this kind and a large number of operators to man and maintain them. R&D to cover a larger range of materials for 3D printing is required. Developing countries in general and India in particular should also orient IOT to their own way of life by applying it less for modifying assembly lines of production than for offering expertise from remote locations for a variety of tasks like preventive or first aid applications in remote areas. Thus, technologies should aim to make small village communities more self-sufficient. 3D printing, AI and IOT are expected to play a much bigger role in more accessible and affordable medicine and surgery in the Indian context. But this requires imaginative handling and a well-defined regulatory framework and mechanisms of dispensing medicines. Combining AI and Big Data can facilitate several decisions requiring discretion in public affairs. It is these types of decisions which are many a time the source of bureaucratic corruption. It is unrealistic to expect these developments happening too soon, but is a goal worth attempting. Even rooftop solar (and other renewables) and micro-grids also fall in this category, where more local people are trained and employed. This minimises the dependence of people living in remote areas on urban centres. In turn, this could minimise migration and lead to major progress. These micro-grids could join bigger grids at a later stage when they are ready to handle fluctuations in load. Smaller projects, in general, are easier to perfect and then multiply in numbers for augmentation of capacity. For developing countries like India, it is important to venture into big projects more judiciously and make use of remotely placed knowledge workers on small projects who have the potential of a large multiplier effect in presently poorly developed areas. This ensures inclusive and sustainable development.

KBE and associated technologies will be more suitable if adapted to domestic techno-socio-economic requirements. If developing countries continue to ape completely the models of developed societies (including China in some cases) in deployment of technologies, it will make them chase a continually moving target. For India, knowledge with appropriate deployment of latest technologies and more importantly developing appropriate human resources and products will be the key to ensure harmonious and inclusive growth. This would demand originality and imagination from researchers, technologists and innovators. KBE without rooting in the domestic context will be a disaster the policymakers must guard against.

INDIAN INNOVATION ECOSYSTEM

Both these words, 'innovation' and ecosystem,' are used quite often while discussing lack of progress of indigenous R&D and its linkage to manufacturing. There is need to understand these words in the context of developing countries. Scientific R&D has always rested on the basis of what is called innovation, but in the present day context, it has acquired new connotation. This has happened particularly after the advent of IT and its role in novel deployment of S&T. Innovation need not be related to science alone, but could also be in mode of deployment or some business model etc. The new context has been enabled by development of ICT equipment like personal computers, mobiles, tablets, techniques of data handling, modes of communication, storage and the phenomena of services aggregation. Very novel means of doing conventional things have caused a paradigm shift. This is distinctive from R&D in physical sciences, the way it used to be related to technology development, building products, developing new processes etc. The subjects of e-marketing, e-governance, e-banking, e-commerce, BPO industries, KPO industries and many more are in the offing. KBC has acquired increasing significance. All such capital like software, designs, marketing, responses to local conditions have become an essential part of innovation ecosystem.

It is quite difficult to define rigorously what constitutes innovation ecosystem. But some examples with diverse backgrounds can offer guidelines. A single university at Colombia, USA could generate dozens of Nobel prizes, a private company Bell Labs could produce seven Nobel prizes and a small geographical area named Silicon Valley is creating a large number of novel enterprises. These offer inspiration, but need not be copied (extremely difficult to copy!). Developing countries need not be despondent at their present-day shortcomings. Most importantly, there is nothing racial about the capability to stimulate such ecosystem. As the present-day America is a melting pot with talent pooled from a large number of countries. Moreover, historically, some of the old civilisations like China, India and Arabs had produced a lot of good innovations in a different time period even though presently they may be relative laggards.

They can regenerate the right ecosystem in modern day context by encouraging freedom for novel ideas, healthy competition, genuine respect for novelty, irrespective of source and a cultural emphasis on creativity in education. The last factor among these is normally underestimated. Presently, many countries give excessive stress on funding, infrastructure and regulatory regimes to sustain innovation and ignore cultural aspects.

Idea to Product Ecology

Ecology is a word with connotations from biology which describes interdependence, struggle and survival of living beings in their natural habitat. The ecosystem for innovation is an apt word because for the world of innovators also there are several players which are all interrelated and the 'habitat' varies from country to country depending upon its level of growth, availability of finance, human capital and policies pursued by the government. The players involved are public sector R&D, private sector R&D, public sector manufacturers, large private industries, MSMEs, startups and various types of consumers or markets. In a biological ecosystem, this interdependence manifests itself in the form of food chains. For example, in a forest, grass is eaten by a grasshopper, which is eaten by a frog, which is eaten by a snake and so on. In a forest, even the corpse of a dead animal is of use to a scavenger (vulture) and remnants are of use to insects or soil. In an innovation ecosystem similar chains rest on human capital, finance and ideas operational in above mentioned players as the starting point to be followed by chain of end users. Such an ecosystem also needs to be equally efficient in use of all kinds of resources.

Encouragement for innovators is dependent also on social, economic and political milieu. Extreme poverty may discourage creativity and route of enterprise for young men coming out of universities. Continued lack of economic security over generations makes them seek jobs instead of harnessing innovation for enterprise as innovation is considered risky. Similarly, bureaucracy in setting up a new enterprise and restrictive regulatory regime in operational aspects is sufficient to dissuade young men of ideas to move on from that track of innovation-based enterprise. Bureaucracy in developing countries, in general, wants to exercise controls by restrictive rules and regulations for the sake of power and pelf. This may discourage many young entrepreneurs from the path of technology-based enterprise. Moderate deprivation on the other hand often may be responsible for stimulating innovation. *Jugaad* (a contrivance born out of necessity) in the Indian context is one good example of this. As India progresses economically, increasing numbers are expected to come out of the shell of economic insecurity and enter the circle of innovation and enterprise.

Normally, the human capital is assimilated by a university or an R&D enterprise, each of these has to link up with an industry which further links up with a market. In this, there are auxiliary chains of say flow of finance and flow of ideas. Finance could come from government funds or private sector funds or even angel investors and venture capital funds. Any scientific research culminating in a marketable product goes through few distinct stages. The first among these rests on basic sciences, to be followed by proof of concept. These two stages are normally in the domain of funding by research institutions. Proof of concept is furthered by related technology development and then a marketable product prototype. The final stage is that of scaled up production. In principle, these last stages can attract funds from private sector industries. These are the industries willing to bet on R&D. Even that is not equally easy in different countries. In India, Industries are not so hopeful about R&D in general, pharmaceuticals and automobile sectors are exceptions. Most difficult stages are the intermediate stages to which both research institutions and industries are indifferent. Nobody feels sure of its successful marketing. In most developed societies, where R&D is easily translated to marketable product, these stages are funded by angel investors or venture capitalists and the like. The most important basis for this is the credibility developed by workers in that area over a long period of time. In the Indian ecosystem, this is the missing link, but recently has seen some action on IT-based services and products. For this nature of funding to diversify to hardware products, Indian R&D has to mature and build a credible face. The

Government of India has recently initiated a fund of funds for young innovation-based startups. Once again, it may be remembered that about 70% of Indian R&D is in public sector which has limited role in promoting manufacturing. Similarly, increasing expenditure of MNC R&D centres is diverted for manufacturing in their home countries. Thus, a very small fraction of investor money is available for innovation-based enterprises.

Innovators and researchers in government sector require attention for harnessing all their efforts. Presently, the government is funding liberally to various departments, but the moot question is, do these funds reach where they should and when they should? This is where bureaucracy of science managers can influence positively or otherwise. Funding mechanisms, committees have to use their discretion with great care. Fairness to genuine innovators is imperative. Many a time, unviable government R&D projects linger on for decades in the hands of incompetent and prove to be white elephants because these were sanctioned on the whims of some 'super boss' and at the same time these funds could be denied to a genuine innovator just because he is the outspoken type or it could be plain perception difference. Many a time, the situation is akin to ‘water, water everywhere and not a drop to drink.’ In general, innovators cannot compete with sycophants and lobbyists. So extra mechanisms need to be in place so that funds reach 'when and where required by an innovator/entrepreneur' — this has to be ensured for the sake of an ideal ecosystem. For large projects, this is applicable to manpower and other required resources as well.

Human capital, its sourcing and flow are one of the most dictating factors. This type of capital is scarce and immobilised presently, while if given proper opportunities it can even appreciate with time and also move between organisations as and where needed. Innovation is demand driven or dream driven and is risky as a business. Though a university can facilitate such innovative creativity, but not all those studying there are capable of it. A university does not impart formal education alone, it prepares mindsets also. In future this will be another role universities and faculty will be expected to play — to inculcate the spirit of market adventure besides creativity.

Real creative talent cannot sit idle and creation is fun for innovators. For both the reasons, innovators will continue to seek pastures which support them best. If there is scarcity of innovators in any locale, the projects for S&T development will suffer. And if there is excess of them, then these talented innovators will get frustrated and will most likely move over. India is in a peculiar situation where a number (innovators and researchers) which should have been in excess is converted to deficit for a variety of reasons. Large numbers of such R&D innovators are moving out of this sector or even out of the national ecosystem. Can one say India is not having enough projects or problems to address? Such a demand is normally articulated by private sector in developed countries, but in India, the government has to travel the extra mile for the same. This outward movement of talent is difficult to curb in the absence of the requisite ecosystem. Korea as a nation has been able to stop this tide of hi-tech migration as many of these workers return after some years of stay in developed countries and thus are of added value. Koreans have also been able to create a virtual network of any such 'lost manpower' and harness it suitably [33]. Even the Chinese have been able to attract back a lot of its talent from foreign shores and deploy them well. Some countries make better use of their 'brain deposit,' a term used recently by the Indian Prime Minister. Despite such statements, India has yet to learn to attract and retain talent in science and technology and harness them usefully. Developing countries need to innovate on projects of relevance (of course for private sector the relevance implies demand and profits) to society and enlist these young researchers to take up relevant challenges before they move on. Developing countries, unlike the developed ones, have not yet succeeded in generating a seamless connect between R&D and market demand.

Flow of Ideas

The second chain is the flow of ideas. Creativity flourishes best in a domain of total freedom — to express, to propose, to interact and collaborate across disciplines, sectors and institutions. For creating the right ecosystem, policymakers have to ensure total freedom regarding ideas as long as they do not impinge on financial or strategic disciplines. Both at national level and international level, one can notice plenty of barriers of various kinds. Within the country, for example, private sector or startups cannot interact easily with researchers from government institutes if they wish to without encountering a lot of bureaucracy. Then there are questions related to proprietary rights and patents. There has to be an environment of trust to facilitate free flow of information and ideas, particularly where commercial interest is important. Internationally, for reasons of commerce, (and pretended strategic/proliferation etc.) there are plenty of technology denial regimes like Non-Proliferation Treaty (NPT), Missile Technology Control Regime (MTCR) and other similar related issues. Indian researchers have to work in many sectors in total isolation. India does not have control over many of them, but wherever it is possible these barriers need minimisation.

Internally there are also cultural barriers which have developed over a short period of India's independent existence. Indian institutions, particularly government R&D, need to promote their credibility not only by proper research, but also by free interactions, outreach and dissemination of ideas, rather than being secretive, possessive and bureaucratic. Indian agricultural research has been most forthcoming in this respect. Many more such possibilities exist at least at national level. Any R&D done on public expenditure is property of the nation and more widely it is used implies greater credit to the innovator. There are several factors at personal and institutional levels which operate to restrict this. Working in silos should be the last thing for government research institutions. Perhaps this may require a fresh intervention framework allowing more free access and flow of ideas and manpower.

Innovators jealously wish to guard their proprietary rights. Thus, it is an important duty of the government to have a fair patent regime. Indian private researchers will be left defenceless against copycats. But suitable checks and balances need to be in place in view of unique position of India vis-à-vis its poor population and lack of patent awareness. It must be realised that there is an onslaught on indigenous R&D and industry by several multinationals from the developed world. A clear example is that of pharma industry. The fight is between providing affordable and proper access of life saving drugs to poor patients and using patent rights to stimulate further innovation. Recently, a United Nations panel has also defended cheap generic medicines for the poor. In many such situations, reverse engineering or frugal engineering is equally a part of innovative process and needs to be duly respected.

Governments have many other tools at their disposal by which to encourage innovation and hi-tech industry. The government, both state and central, are the biggest spenders on several sectors like railways, defence, aerospace, telecom and so on. India must have some sort of incentivised means of procurement for products and services born out of process of indigenous innovation. This has been recommended several times by several advisory groups, but is not followed in practice as this contradicts the practice of precedents and lowest quotation followed often by the bureaucracy. Similarly, claims of innovation-based industries were rejected in large numbers as not fitting the category of startups. The government and bureaucracy handling these efforts have to be more responsive and dynamic. There is a feeling among several working level innovators that governments create hurdles because things without a precedent are not respected by bureaucracy and later on may be a subject of audit queries. It is fair to demand background check, but there is need to follow best international practices favoring the innovators, particularly small players. Even a small level innovator should be able to approach a kind of proposed ombudsman for hearings related to complaints of unfair treatment. If a local supplier is able to indigenise a product for a public sector institution, then he should be allowed to sell that product recovering at least 90% of the cost at which it might have been imported — if it fulfills the quality criteria. Fair competition with respect to bigger private and government players is not an unreasonable demand, particularly when in the near future there are likely to be many new startups.

World over, different yardsticks have been identified to measure the level of innovation growth and the associated ecosystem. Some of the parameters considered for this purpose are expenditure on R&D, quality of research institutions, availability of quality and number of scientists and technologists, university industry collaboration and number of patents [44]. From among these, India fairs reasonably well on funding and to some extent on availability of number (not quality) of scientists and technologists, but poorly on three parameters, viz. (i) quality of institutions, (ii) industry university collaboration and (iii) number of patents — particularly those worthy of commercialisation. As can be seen, building up quality institutions is dependent on talent in them and requires sustained efforts and time, while collaborations between industry and universities is related to the attitude and mindset of both the parties involved. This could be partly prevailing culture, inertia, or related to credibility deficit between the two. This has been repeatedly emphasised during the course of this work. Even patents are outcome of applied or industrial research and have links to attitude and incentives. For example, a large fraction of government researchers builds up its career with no requirements on patents. This must be contrasted with the case of China, where the government offers huge incentives several times the salary for papers published in prestigious journals or worthwhile patents. If the Indian government is really keen to have the right ecosystem for innovation, then there is a strong need for encouraging change of mindsets. About world-class institutions also there is a rider. While international prescriptions will harp on that, but India has to carefully balance between creating world-class institutions and retention of talent produced in them. As it is often seen that talent in world-class institutions, trained at government expense by developing countries is lured away by attractive salary and working conditions in developed countries.

While America, Japan and some European countries have well established starting point in terms of growth on patents, among newcomers, noticeable progress has been made by China. Other BRICS countries have also progressed, but not to the extent of China. India presently ranks 57th on Global Innovation Index, more than 30 notches below China. Whatever startups euphoria exists in India at the moment is more related to IT sector, ecommerce etc. and even there, many of the ideas seem to be copy of what has been already implemented in developed countries. India can continue to encourage creativity in IT-based products but much greater emphasis on hardware-based startups and innovations is required for the sake of a balanced ecosystem.

INDIAN NATIONAL TECHNOLOGY REPOSITORIES

In a vast country like India, there exists huge data, which may be required for any worthwhile analytical consideration for policy inputs. This makes the task very daunting in general. This has led to a culture of indifference or halfhearted collection of several kinds of data leading to inadequate policy/action response. For several purposes there is a tendency to depend upon foreign institutions or even multinational consultancy groups for purely domestic data. In that respect, development of database of unique biometric identification system of Indian citizens (covering more than billions of its population), also called 'Aadhaar,' is genuinely path breaking for the sheer magnitude of the exercise as well as its multifarious implications. Another similar exercise, carried out recently, has been related to handling a single taxation system called GSTN (Goods & Services Tax Network). Both these exercises indicate the level of competence generated domestically in the context of large data handling boding well for the future.

When one considers need for an analogous comprehensive data collection in the context of STEM HE, scientific research, technology development and manufacturing industries, it may be realised that the existing mechanisms for this purpose may not suffice and may need a radically different approach and methodology. Such an exercise is required for the sake of developing a robust database and generating more specific action points for a progressive STEM HE, research, technology development and manufacturing ecosystem. Such database can have positive implications for appropriate policy responses, constructing adequate support structures, identifying gap areas, promoting collaborations and synergy, avoiding unnecessary competition and many more such benefits.

The potential of such database needs to be explored further and it seems possibilities are immense. Such database can focus on institutions, organisations, industries, human resources and technologies. A recent indigenous database in the context of higher education has been the rankings of HEIs in the country. So far, India had depended on global rankings and not many HEIs had figured on those listings. Thus, patchy information about few institutions does not serve desirable purpose. The HRD ministry's initiative in this regard is a useful beginning, but needs some localisation by proper choice of weightage factors to different parameters. In the context of present discussion aimed at STEM HE and R&D for the mission 'Made in India,' the type of database which can be much more useful is the one which is more specific about individual STEM departments. If all STEM HE departments in the country submit their relevant data to the ranking agency, the ranks decided can be a great boon to all stakeholders like aspiring students, researchers, faculty, or prospective employers. These rankings can be decided on annual basis. Modern methods of data collection and authentication need to be employed and data inputs from these departments should be mandatory. Once all the data as well as rankings are publicised, it may stimulate competition among the participants, offer students a range of better defined choices and also give employing institution/organisation or industry a clear picture of the background of the recruit.

The second category of database of concern is that related to human resources in STEM subjects. India has a very large number of students in STEM subjects and the quantity is taking toll on quality. Student output from few elite institutions is insufficient and the problem is aggravated by poor retention of talent within the ecosystem or even within the country. The real 'guts' of the Indian system lie in its numbers and if good care is taken of them, they can prove to be national assets, or otherwise they can turn into a demographic disaster. Students in remote or small universities/colleges need good mentoring and interactions with inspiring role models, assurances about good career profile in STEM/industry ecosystem. All these need quality assurance of students and researchers, monitoring their loss/migration from the ecosystem or fulfilling the mentoring needs. This can be achieved through a comprehensive database of STEM HE human resources in the country. Such database has to be dynamic with regular updates. It is obvious the details expected from such objectives are not readily available and new machinery has to be activated. An associated requirement is about skills in STEM subjects with implications for manufacturing industries. OECD countries have a skillset management system which helps all stakeholders with information to manage a balance between supply and demand of skills. Such database is need of the hour for India as movement from general talk about skills to specific action points should be carefully crafted. Such database of skills and their demand is required by students, employers and policymakers. As a country evolves, its needs change and this database of supply and demand should reflect this.

The final pillar of the 'Made in India' mission is technology and industry. Present day database of technologies exists in silos of different departments and DST. In most cases, it is not used to initiate as well as stimulate new technology development missions — filling the existing gaps and envisioning futuristic ones. This way, the data loses its potential to encourage technological progress. Similarly, data about industries is underutilised. This exists mostly in government records as a book keeping exercise more related to commerce than to promote technology's role for industry. For example, a very large fraction of India's industry is in the form of MSMEs. Recent operationalisation of GSTN indicated 9.5 million MSME units. This huge database can be further made use of in variety of ways.

Indian MSME units are very poor in terms of productivity and assimilation of new technologies and the government keeps subsidising them in context of so called technology upgradation. This mostly amounts to importing new machinery from foreign countries. Thus, indirectly, India keeps subsidising development of technology in those countries. If the entire data of these 9.5 million industrial units is analysed and categorised sector wise and their technology needs are identified by 'specialised technology scout groups,' then it may turn out that basic capability for 'technology upgradation' of most of them may be existing within the country. As can be seen, the starting point is a well analysed database. Besides technology upgradation, a culture of assimilation and adaptation of new technologies may be inculcated in the process in these MSMEs, which are in fact the backbone of Indian industry.

Technology trade data of a country is very useful information for identifying its status of technology development and gap areas. This trade happens in two forms — pure trade of embodied technological products and diffusion of disembodied technology in the form of transfer of techniques, products and services employing patents, designs, knowhow and licensing. Economists often employ what is called technology balance of payments/trade as a concept to understand the extent of benefit or loss related to technology export/import. Secondly, this could also be used by those responsible for development of technology in the country. The difference in the two cases is that second data demands detailed information of products with components, materials and sub-systems with specifications. For developed countries, a surplus in this balance indicates the extent of technology development. For example, for the year 2011, this surplus was $121 billion for USA, $54 billion for EU nations, $29 billion for Japan. Korea recorded the largest amount of trade deficits in electronics sector despite the fact that this sector contributed maximum to its exports [33]. The reason being import of components and export of products not being rightly balanced. Such sectoral database can be usefully employed for stimulating growth in the deficit area.

Such database on technologies can also be employed for those working in strategic sector by identifying lists of critical components/technologies, likely offset areas and managing dual use technologies. Such data can also help identify the rate of assimilation of a particular technology. For each specific case response mechanism can be decided upon in an efficient manner in consultation with other R&D or industrial entities.

MIGRATORY BIRDS

In biology, it was an amazing find that birds migrate when they find clime and conditions not so conducive to comfortable living. So as winter approaches, a large number of such birds from Siberia start flying towards areas around lakes in the Indian subcontinent. There are a host of other phenomena about them which are interesting in themselves, but here we are concerned with the chief fact that they migrate when conditions are not conducive for their survival.

Come September and on several airports across India, one can see hordes of young men and women catching westward flights. These are all bright and brilliant youngsters who purportedly are leaving India for higher education. They leave behind their families and motherland seeking a bright future for themselves in foreign lands. What was supposed to be "higher education" later turns out to be a 'honeytrap' of migration. This phenomenon is not restricted to India alone and many Asian countries are affected by it, though each is affected differently.

Over the last few decades, there have been different shades of this migration and the perceptions about it have also changed with time. In the sixties, it was restricted to some bright graduates from the IITs who went for higher education and making their mark in international science and technology. Slowly, science students from some of the bigger universities too joined the bandwagon. Then in the late seventies after the Gulf boom, many of the skilled workers left for Middle East. These were mostly technicians and artisans. It is noticeable that science and engineering graduates could not find place there at that time, but recently even for them, demand conditions are improving. So, there is a source country which has not much to offer in contrast to a destination country which has plenty to offer in terms of higher education, immediate salary hike, better living conditions, better opportunities for growth and so on. For some students of S&T, much more important is freedom and to be able to work with top ranking scientists on frontier problems. They feel this provides them the opportunity to show their mettle and excel.

It is significant how Indian policymakers and society viewed these migrations over the years. In the first place, these were treated as some kind of 'brain drain.' Some noises were made by the interested parties, but nothing substantive emerged This was the period when private industry was not so much in the upswing in India and which heavily relied on licensed production models and R&D and indigenous innovation was not much valued by them. Then in the second phase when the country was going through extreme foreign exchange crisis, many policymakers counted the dollars of remittances from them for alleviating a national economic problem. Presently, India is the biggest receiver of such remittances globally, even though in their student phase these migrants depend upon their parents back home. In all these years, this migration had been growing in magnitude. What was a trickle then is a deluge now. There are hordes of them leaving every year. What was cloaked earlier as 'higher education' is now obviously a shift for better jobs and careers. The seeds sown earlier are fructifying now as one sees the likes of CEOs of Google, Microsoft, Nokia, Pepsi, MasterCard and many others as Indians, who left India two or more decades earlier. But what is more important is not these symbolic heads of a few MNCs, but a much larger number of lower and middle rungs of S&T workers who are working in these foreign S&T and manufacturing organisations and are a loss to the nation.

Whatever be the achievement of individuals, collectively they have made good impressions. About 16% of startups in the Silicon Valley are by Indians. Just like from 'population problem' to demographic dividend for the case of increasing population, it has been a long journey to treat this problem of migration from prism of 'brain drain' to remittances to what is now called 'skill deficit,' but also being portrayed as 'brain trust' to be exploited later. The fast-changing scenario in technology development calls for new age skills. The poor growth of many S&T-related sectors of R&D and manufacturing have been responsible for depriving skilled S&T workers of the opportunity to grow in India and seek greener pastures abroad. Poor record on growth of socio-economic scene has also aggravated the problem.

There is need to examine why these young and bright minds prefer to migrate and what India and developing countries in general have not done to retain them at home. This poses complex questions which go beyond the choice between pragmatism and loyalty to the nation. India is a middle ranking country with burgeoning middle class. There is a fast-changing world with a lot of progress in S&T field. It is natural that many of these bright youngsters will like to share the glory of international science and technology. If at the same time it helps them solve their personal economic problems or gain in social stature for their families, they feel it to be a goal worth pursuing. This is all the more so when they find that by living here they can serve the purpose neither of Indian progress nor of international science and technology. The aim is to dispassionately understand and appreciate those who have left for the sake of personal goals. The price of emigration that is paid by those who migrate is to be forced to live in a culturally alien environment and initial euphoria of fat pay packets and glory of working in international science and technology system does not sustain, but they get used to their new way of life and find it difficult to return also for fear of non-assimilation. If the Indian S&T system, industries and its socio-economic system can generate conditions conducive enough, this deluge of migration can be arrested, if not completely stopped.

Having touched upon ‘how and why’ of individual perspectives of migrating S&T workers, the next issue is what nations and their science and technology system can do to alleviate the migration of talent. This must be viewed in comparison to what developed countries are doing to attract talent.

There have been several studies that analyse the impact of immigration. The focus is on high skill migration, but it considers medium level skills too. It recommends for host countries thus — transforming the employment level of migrants similar to that of natives can raise significant economic returns for that country. EU Blue Card Directive is an initiative in this direction. In response to this recommendation, at least seven OECD countries have brought major reforms in their immigration system for skilled labour. They hope such reforms will bring them to the centre stage of competition for global talent. They encourage foreign migrants in S&T at the cost of domestic talent and aim to provide citizenship to these talented lots as early as possible to avoid these people returning to their native countries and give competition to developed nation. The recent noises about H1B visas in USA are to regulate such migrations further to their advantage. Some of the protectionist spirit in many other countries should not be mistaken that they are giving up on hi-tech migrants, but these atmospherics must be used by India and other developing countries to attract and bring back many such high worth scientists and technologists. It was Dr. Homi Bhabha who worked hard to bring back from foreign lands many bright scientists and technologists and find suitable employment for them. In fact, it is reported that just before his death, he recommended the name of Prof. Chandrashekhar (who later won the Nobel Prize) as his successor for the post of Tata Institute of Fundamental Research (TIFR) director. These were isolated cases. It seems, as a nation, many of these countries do not have any policy to attract and retain talent and prevent migration. In contrast, China has launched major programmes to attract back Chinese talent from America, with fair success.

The difficult question to answer is what can be done. First of all, is the need to acknowledge it as a problem? There is a feeling among many that such migration reduces India's headache in terms of providing jobs. At different levels, many science managers may be trying on their own to act in individual cases, but that cannot be a solution to this problem. Systemic response is a must. Action is required on several fronts and it is a problem which cannot disappear soon. The most important step to be able to attract talent is to reform S&T system, institutes of higher education and cultivate/accelerate R&D in industrial sector as that can attract more capital into the ecosystem, which can then attract talent back by offering large size projects as well as salaries. The socio-economic conditions of S&T workers must be improved with transparency in funding for R&D projects with due incentives for enterprise in hi-tech industries for first generation entrepreneurs. Recently, a lot of hue and cry was made about the flight of few millionaires from India for reasons of probable tax terror.

But talent is being lost to the country for several years, but nobody raises voice. It speaks a lot about the mindset of the nation.

While all these efforts take time, one should not forget to leverage these human assets who have Indian roots. First of all, there has to be a system of keeping track of all of STEM talent that leaves Indian shores and the database about their activities needs updating. India should aim to harness them even if they do not want to return. Most of them will gladly join this exercise. The first aim should be that whenever the conditions are ripe for an individual to return, he should be encouraged for the same. This should be decided mutually, say by offering a suitable opening. This centrally processed data about skilled non-resident Indians (NRIs) should be accessible to local industry, entrepreneurs and S&T institutes. Sector wise, there should be virtual networking between people at both ends. Even linked institutes or industries should be in touch through this database and virtual network with skilled personnel of their own field. This networking/database can serve multiple purposes. First of all, those working in Hi-tech areas can flag about latest technological advances (without compromising proprietary information) and stimulate local R&D. Secondly, they could be part of a network of experts/specialists who can be on board of assessment for high level (multiple levels) posts/appointments and project appraisals. It is obvious that there will be strong resistance to this from local entrenched "experts" and other vested interests. But India as a nation must weigh and realise that this could provide a fresh whiff devoid of prejudice. It must be remembered that there is often acute shortage of the right kind of experts in several areas. Additionally, these NRI contributors could, in principle, be relatively more fair and unbiased referees (with careful choice). The fee charged by them should not be a discouragement and should be weighed against benefits it will bring by way of facilitating the right choices. Much of this work can be facilitated through modern means of communication, including video conferencing.

FOREIGN TECHNOLOGY AND TECHNOLOGICAL INDIAN

India gained independence from British rule in 1947. As usually happens with colonies, there was too little scientific research and even less manufacturing within India. It was said metaphorically that even a sewing needle was imported from Britain, before independence. Only after independence, the new government started laying stress on indigenous R&D as well as manufacturing. As part of its independence struggle, there had been a political movement to favour indigenous products — the *Swadeshi Movement* (pre-independence Indian movement to buy only indigenous products over products imported from Britain). This movement aimed at boycotting 'Phoren' (local colloquial for foreign) products. After attaining independence, less need was felt to boycott imported technology. To this day, respect for quality of products imported from foreign countries does exist. Locally manufactured products in India have never acquired greater respect against imported products in terms of quality. A large number of Indian companies name their products in an anglicised terminology to lure gullible Indian customers, who believe it to be of some foreign connection. Excessive focus of Indian manufacturers on low price at the cost of quality resulted in a craze, among vast sections of its population, for 'phoren technology' (foreign technology products). This is applicable as much to lay consumers as to researchers buying equipment for their products or the armed forces buying weapon systems. This continues unabated to this day, despite the fact that India now produces a vast range of high technology products like nuclear reactors, spacecrafts and fighter aircrafts.

Just after independence, there were almost no institutes of STEM HE in India. There was a great push by the Indian government to nurture high technology within the country. India established a string of Indian IITs in various parts of the country. These were on the lines of MIT of USA and were aimed to be of world class. Soon, these institutes started producing scientists and technologists of high quality. These technologists found themselves as if on an island of high technology amidst a sea of ignorance and poverty in the country. A large majority of them felt they may be more useful in the developed world. A wave of migration from these institutes started to countries like America. This wave was further fueled by personal ambitions. These products of IITs found hearty welcome in America, the land of opportunity. They built up a global brand for themselves and networked among themselves with great success and contributed to global technology ecosystem. They proved themselves to be the backbone of many high technology companies and startups in America in general and Silicon Valley in particular. These technologist settlers from IITs joined the league of a large number of prominent Indian technologists who came to study in MIT earlier and decided to settle there. It is about this latter category that Ross Basset talks about in his book 'Technological Indian' and calls them as technological Indians [46]. He points out how many of these made a worthwhile impact on major global technological developments and contributed to rise of America as a technology power. These developments can be viewed together with the fact that indigenous technologies on various fronts were struggling to establish themselves, Indian industries were banking on licensed technologies for manufacturing and Indian consumers continued to crave for 'phoren' technology. Ross Basset in his book, *Technological Indian,* has chronicled all the Indian technologists who graduated from MIT up to year 2000 along with their contributions to the world of technology. He comments about how they failed to contribute to growth of technology and hi-tech manufacturing in their own country.

Every year, lakhs of Indian students strive to enter these esteemed institutes and Indians feel proud of the contributions of technologists coming out of these institutions. But this seems to strengthen the craze for 'phoren' technologies, products as well as degrees (or even education system). One may say that having a world-class institution in a developing country is a double edged sword. Unless there is a critical mass of researchers in a technology sector within the country, it may not be possible to stem the tide of their migration out of the country. It is a vicious trap and there is need for the government to pay greater attention to retaining talent than to setting up world-class institutions. At the same time, Indian society must learn to respect and value whatever indigenous technologies are developed with totally Indian efforts. One can only wish spinoffs from Chandrayaan and Mangalyaan (Indian missions to moon and Mars) are available on Indian supermarket shelves.

WOMEN IN STEM ECOSYSTEM IN INDIA

Gender equality is a global issue at present. Social scientists and women activists had been advocating it for a long time, but now economists have started estimating the price that a nation pays for such inequality. But in India and many other developing countries, these inequalities reach a level of criminality, except for some of the bigger cities where women compete fairly well with their male counterparts. Also, there are several exceptions where women have entered the bastions reserved for males. They also excel in several competitive examinations beating men. This still does not indicate the true picture of women's limited role as participants in the progress of Indian society. These facts become more pronounced in the case of higher education in general and STEM HE in particular.

This is a cultural lacuna which needs proper understanding so that it can be addressed suitably. In traditional Indian society, women are not expected to go for higher education and STEM subjects are considered to be exclusive domain of males, with limited participation from women. There are additional reasons which limit their participation in these subjects. The employment potential of these subjects has the probability of their landing in manufacturing plants (with maybe night shift!) affecting their family life. Possibilities of an unfavourable geographical location, odd working hours, facing male dominated environment and family responsibilities are some of the factors which discourage them. For example, among various IITs, the percentage of women in IITs in Mumbai, Delhi and Mandi are higher as compared to Guwahati and Kanpur. Perhaps perceptions of remoteness from mainland and law and order situation could be part of the reason. Recently, the Government of India has realised part of the problem and initiated programmes to make young girls interested in

STEM subjects. The government has also asked all IITs to increase recruitment of girls by going beyond their normal share by the merit list. It must be realised that IITs cover a miniscule fraction of Indian student force. Additionally, entry into educational institutions is only part of the story. There have to be assurances on employment and its nature too. Increasing role of softer jobs in futuristic knowledge-based economy and possibilities of working from remote locations could be advantageously employed to facilitate entry of women in STEM related jobs.

What is the role of women in STEM R&D ecosystem and how various constraints and factors play a part? According to DST, as on April 2015, there were about 5.3 lakh personnel engaged in R&D establishments (including in house private and public sector institutions) in India. Of these, about 2.8 lakh were engaged in active R&D and about 39000 of them were women (about 13%) [5]. This ratio is quite poor and compares very unfavourably to the average in developed countries. Active encouragement steps by the government and proactive soliciting outreach by R&D institutions can have big potential impact. Indian institutions in STEM R&D and higher education have to stress upon the fact that their teaching and research jobs offer good amount of stability. This may be in contrast to manufacturing jobs. The impression on mind of young female students about STEM HE leading to manufacturing jobs could be suitably deflected. In fact, Indian IT sector has already exploited this to some extent. This has to be viewed together with the fact that there is greater probability of young women in STEM R&D not migrating out of the country for reasons of family responsibilities. This could have a favourable impact on retention of talent within the Indian ecosystem. India has to create an environment where women contribute actively to say, softer side of knowledge-based capital with incentives to work from home and more freedom on having break in service in case of genuine family requirements.

Innovation Friendly Management

I have seen many phases of life; I have been minister of state but if I had to live my life again, I would always remain in my lab, for the greatest joy of my life has been to accomplish original scientific work.
Jean Baptiste- Andre Dumas

Indian scientific or industrial establishments of all kinds need to give highest importance to innovation and enterprise. Bringing up such a value system depends heavily on the type of leadership in these institutions. In Indian scientific research labs, three operational styles of leadership commonly observed are either coercive or authoritative, consensus-based and *guru-shishya* (intimate relationship between teacher and pupil) mode. The last one owes itself to deep rooted traditions of Indian society. The first one is more repugnant to creative types but is quite often employed in times of pressure or in regimented projects. In the Indian environment, the last mode is most appealing, particularly in universities, and is many times successful. But this is not without disadvantages, as it fuels groupism (or lobbies). Many of Indian science and engineering academies have good fraction of membership whose lineage can be traced back to *shishya* following *guru*. This is not to belittle the worthiness of any of them. In this age of innovation and knowledge-based economies, it is very important to be able to stimulate innovative environment and any management style that discourages innovation and competition has to be curbed. In India, more than half of funding (and correspondingly encouragement) is cornered by big national labs, so healthy science management style is imperative in these labs for the sake of spreading such a culture in STEM R&D/industry ecosystem. It must be clear that it is not good enough to be able to attract talent (though even there many institutions could be lacking) but to retain it and provide it the right environment is challenging. For many of these institutes, delivery on mandate is pressing, which drives them to coerce and regiment, but these provide short-term gains at the cost of long-term benefits. This aspect of the problem requires more serious attention from policy makers at several levels. What kind of science managers India needs should be debated. There are no universal solutions. Many of the science managers are lured by positions of power and instead of consolidating and expanding their own field of research, they start concentrating too early in their careers on diversifying and expanding beyond their own turf. Ambition of a scientist has to be healthy — a balance between individual and national needs.

This is all about ambitions and attractions of positions of power versus the passion to be a scientist. Many times, there is a conflict faced by scientists in choice of being in leadership position or continuing to pursue science. Thus, he has to decide between his own area of research work, its consolidation and strengthening on one hand and diversifying to unchartered territories, on the other. If he makes the later choice, he exercises power and pelf, but he starts getting away from true scientific work as is known to the scientists. Sometimes of course, it is beyond his control and he genuinely transform to being a leader, but more often he falls into a trap and is reduced to being only a boss. As Lawrence, father of particle accelerators, once mentioned, the net worth of a scientist does not reside in his technical skills alone, but also his human qualities. As it happens, larger numbers do turn into being bosses and move away from their actual strong point of being scientist. If these situations are not handled carefully by top level science managers, it is a loss to the world of science and also making larger number of bosses instead of leaders. Once such wrong choices are made, they take sufficiently long time to remedy themselves. These wrong choices are made either because of poor judgement at the level of top leadership or are related to sycophancy. If the choice is wrong and not based on performance and past track record, then some among the peer group are likely to be demoralised, which is an additional loss.

If some kind of groupism creeps in decision-making process, mistakes are likely to happen. This group favour (or bias) could be related to subject or more human at the level of regionalism or caste. An additional aspect is related to lack of transparency in this decision-making process which complicates matters further. The best that a scientific organisation can do under the circumstances is to minimise the attraction for positions of power and reduce the layers of hierarchy. In India different scientific institutions could be rated differently on these yardsticks. There is usually no formal training for other skills expected from R&D leader. The required skills are to be able to motivate junior scientists, to be able to communicate and articulate about various aspects of his (or his organisation's) work and to be able to strategise the course of action and advise administrative or political leadership suitably. It is said that greatest impediment to development of R&D leaders with administrative, business and strategic acumen is, R&D culture itself.

If one examines the pattern of leadership in many of Indian institutions, it is quite often that group leaders or even heads of organisations are appointed in some sort of 'hereditary' pattern. What is meant is that successor often has the same perspective or belongs to same school of thought as his predecessor. The perspective in question is related to scientific or technological worldview. Such situation sometimes continues for decades together. The organisation is thus deprived of benefits of alternate perspectives and ideas. Close to the end of scientific careers, individuals behave like runners in a relay race and wish to pass the baton to one of their own. But for the sake of national interest, it is best if the baton is passed in the most fair and balanced manner. Leadership selection in scientific institutions in Indian universities and R&D labs has not been given as much attention as it deserves and having a leader which values innovation the maximum is given even less weightage. Following administrative principles of seniority and similar attributes is found safer option to discourage arbitrariness. In India the total process is yet to acquire maturity that is expected.

As can be seen from above description of some of the cultural issues, there are major demands on most players, be it the respect for innovation, the way research directions are chosen, the way research results are harnessed, and the way research is funded. Furthermore, the inability to look at affirmative action beyond the socio-political prism, the inability to look at STEM HE as a strategic necessity, respecting only major indigenous technology projects like Mangalyaan (Indian mission to MARS), but ignoring homemade high technology products in the market, and the inability of researchers and industrialists to join hands are other major cultural issues at play.

Chapter 8

CONCLUSION: JOINING HANDS

This chapter aims to summarise the earlier chapters, covering diverse topics and players in the effort to improve India's prosperity. In recent times, scientific research has stimulated globally, the development of a host of radically novel technologies which have the potential to influence lifestyles and livelihood of millions of people the world over. These technologies based on AI, IOT, robotics, additive manufacturing, nanotechnology etc. may have a major impact on global manufacturing and consequently the employment scenario. In the last few decades, the share of global employment based on manufacturing has declined, but sectors related to high technologies have not been affected. Some developing countries have gained because of outsourcing of low technology tasks to them. Some of the developed countries that have banked on scientific research and high technology-based manufacturing have not suffered much in terms of employment despite outsourcing their low technology tasks. But these new technologies are predicted to have a major negative influence on employment. Businesses and other entities involved with these technologies continue to discount such possibilities, citing possibilities of new kind of jobs in future. Manufacturing based on these technologies may demand higher level of knowledge base and skills. This poses a major challenge to developing countries with a large fraction of population being poor and unskilled.

Presently, the ownership structure of most high technologies is such that more than 90% of patents are in the hands of a small number of nations and their corporations. It is very likely that these entities will also usher the world into manufacturing based on these novel and disruptive technologies. The gap between developed and developing countries, in all probability, may widen. An employment crisis may also affect many developing countries unless they respond in a befitting manner. Various countries have been responding differently to the evolving scene of technology-based manufacturing. Many of them have made rapid advances in the last few decades and countries like Japan, Korea, Taiwan and China have catapulted themselves into the probable position of competitor to developed nations. These countries have employed various tools like licensing technology, reverse engineering, FDI, strengthening tertiary education system and scientific research and technology development for this purpose. There is a set of Asian countries that have attracted manufacturers to their shores by favourable regulatory regimes, low wages and robust infrastructure. But new technologies on the horizon suggest that the complexion may not remain the same for long. Only the countries with prowess in these technologies may benefit most.

India and many such developing countries continue to be exporting traditional goods, commodities and low and medium technology products. Most of them are caught in vicious cycles which drag down their technology and knowledge-based growth. Low technology products offer poor returns with prices dictated by competition in global markets and WTO. On the other hand, high technology products, particularly those based on patents, are highly priced and dictated by the supplier. This earns them sufficient returns to invest further in R&D — a virtuous cycle for hi-tech products. In countries with large poor population, surplus of labour, low purchasing power and unskilled workforce, there is little respect for efficient means of production. This leads to technicians being employed as engineers, while good quality engineers do not find enough attractive avenues and those belonging to natural sciences are totally squeezed out of the cycle of innovative contributions to industrial manufacturing. This results in industries deriving their competitive strength less from innovation than other means. This leads to loss of talented researchers and technologists to global ecosystem. Another drag is from the quality and kind of research. The research careers in most of these countries seek certification from international authorities. This makes them look for pastures in the international frontline. But mostly these researchers end up in the 'also ran' category rather than genuine frontline, without much benefit to their domestic ecosystem. In the process, originality of research directions rooted to local necessities is the causality.

Though a large number of developing countries has similar problems in the context of technology and innovation-based manufacturing, this work focusses primarily on the Indian ecosystem. India has dormant potential in this context and can be a role model for many other developing countries. Presently, it has a rapidly growing service sector and there are lot of policymakers who feel that greater emphasis on 'softer' technologies, with presumably lower gestation periods is preferable compared to that in hi-tech manufacturing. The incumbent government has launched a mission of 'Make in India.' It mainly banks on manufacturing within India, but there is little or no emphasis on manufacturing based on indigenous scientific research, technology development and innovation, and the yardstick of progress is the magnitude of FDI.

What restricts India's potential to blossom is not simply its scientific and technology ecosystem. The barriers are like a multi headed hydra which requires to be handled at the highest level. There are several structural, policy as well as cultural issues at stake. Even seventy years after independence, the government is spending a whopping 70% of the total national R&D expenditure, bureaucracy in government funded R&D has not reduced, mandates are rigid and monopolistic, limiting innovation and leading to extremely limited connect between university research and industrial manufacturing. There is an aspiration to have world-class institutions, but little attention is being paid to attracting and retaining world-class talent. Further, there is low respect for genuine merit and preference is given to paper culture instead of patents and products. There is also contempt for technology-based enterprise as against a 'decent' government job, shortage of good quality faculty, unbalanced approach in education between concepts and practical education, culture of preferring to develop big systems rather than small foundational components and several others. Similarly, there are cultural and attitude-related issues in plenty which harm the overall growth, isolated and individual achievements notwithstanding. India has a large number of institutions of STEM HE, but only a small fraction is involved in R&D or technology development and directions of R&D are dictated by legacies or limitations of infrastructure or mentorships. Policymakers enhancing funding marginally as an annual ritual or offering a few tax sops occasionally to private researchers or startups, believe it to be sufficient. Scientific managers prefer safety by focussing on their mandates and often ignore offbeat innovations happening under their nose. Private R&D managers treat R&D less as part of strategic vision and more as tactical tinkering. Even fresh startups prefer softer sectors.

The above analysis offers a glimpse of the Indian S&T ecosystem and its linkage to manufacturing. Expecting it to evolve to a state where STEM R&D is linked intimately to the ecosystem of designed, developed & Made in India may seem a tall order. Such an ecosystem may not be able to respond effectively to global challenges of disruptive technologies. Technologies today are changing fast and there has to be a fast and dynamic response by nations to be competitive. Who can be held accountable for this? Several innovative technologies may not fall in the domain of the existing mandates of most institutions. There are a large number of government R&D labs, scientific, technological and industry-linked departments and even ministries. Similarly, there are several high powered advisory bodies. This multitude of organisations often has an effect of causing lethargy in decision making — slow, disjointed and fragmented response and poor accountability of the process as a whole. In view of this, there is a strong need for a mechanism which can coordinate well between various players, which has full authority and can be held accountable in many ways. This may be called the National Science Technology and Innovation Authority or System (NSTIA or NSTIS). This may comprise of eminent scientists, technologists, industrial research leaders, STEM educationists and representatives from the startup community. It may co-opt support from academies of science and engineering as well chambers of commerce. Its main objective will be to oversee reform and reinvention of an ecosystem of STEM higher education, R&D sector (both public and private), hi-tech manufacturing, STEM talent management and a suitable symbiotic environment between various players involved in science, technology and manufacturing. It has to assume multiple roles as per demands — generating databases on human resources and technologies and competencies, monitoring generation and flow of technologies in the country, assuring its widest possible dissemination, organising scout groups for matchmaking related to technology collaborations and flow, ensuring best value for India's natural resources, curbing unhealthy practices in institutes of STEM HE, organising mentor groups for various levels of STEM workforce, optimising supply and demand of STEM workforce and many such similar tasks. This body may take assistance from all existing structures for such purposes. This body may monitor at the macro level and prod the policymakers on one hand and executive entity on the other. Some of these tasks are described below in brief. This description is divided into three parts, viz. NSTIS, STEM R&D ecosystem and few suggested R&D themes applicable to most sectors.

Spearheading Reforms in STEM HE

STEM HE is the fountainhead of all research, technology development and manufacturing, and impacts most of these activities related to strategic sectors as well. For these reasons as well as facing the onslaught of new and disruptive technologies, it is imperative that India declares STEM HE as one of its strategic objectives and segregates it from the rest of higher education. NSTIS has to take the lead in achieving this goal. The first and foremost requirement in this context is to increase the level of autonomy for all STEM departments and institutions to collaborate with academic and industrial partners, sharing of facilities, human resources as well as proprietary rights between partners. To avoid misuse of this autonomy, all these STEM departments/institutions need to be ranked so that stakeholders are not misled. NSTIS also needs to evolve fresh norms and guidelines for selection of leadership in these departments. The process should be transparent, fair, objective and in sync with national science and technology goals. All professor or higher level of appointments may be vetted suitably by a national level body set up by NSTIS and all such appointments should be outside the home state for a minimum stipulated period. This is required to discourage the growth of inbreeding and region oriented lobbies at the cost of genuine merit. Among the major reforms required to reorient the R&D directions and their link with career growth is to ensure a balanced growth of papers, patents and products in all institutes of higher education. Excessive weightage to papers even in applied and experimental sciences, without much quality, may need to be gradually phased out. For the sake of promoting quality, suitable national accreditation system for all STEM theses may be set up by NSTIS with the aim to phase out low quality theses and mentors. NSTIS also needs to ensure that the government maintains a balance in support between elite and mass institutions. This is not purely an egalitarian exercise, but one to make talent search more broad-based. In many of these mass institutions, there are deserving young persons who may need to be exposed to suitable mentoring and role models. NSTIS may have to organise a system of mentors, drawn from a variety of sources for such talented youngsters. The list of desirable reforms expected from NSTIS could be longer still.

Revisiting Mandates of Major Government R&D Labs and Public Sector Manufacturers

There are a large number of government R&D labs which were set up several decades ago. Also, there were several labs meant for auxiliary services like standards, testing facilities etc. The mandates of most of them were defined in the technology context of the day. With new challenges emerging on the horizon, it may be worthwhile to reframe these mandates freshly. This could be done in the context of a new holistic vision that NSTIS evolves for the sake of thrust on technology-based manufacturing and meeting the challenges of new disruptive technologies. The Government of India set up a large number of manufacturing units, also called PSUs, in the fifties and sixties of the last century. Many of them have been making losses for a long time. Most of the time, their functioning has been reviewed — mostly from a commercial angle, but rarely from technology point of view. It should be the task of NSTIS to reorient and restructure them from technology angle and suitably link them to scientific research within the country, irrespective of whether these are privatised or not.

Managing Flow of Technologies

There are two types of technology flow which are pertinent here. This flow could be in the form of licenses, copyright transfers (and other variations) or technology embodied in the form of specialised products. Further, this flow could be either international or domestic. The task for NSTIS is to identify the needs of high technology development, deployment and dissemination within the country. Sector wise technology balance of trade (imports vs exports) is a good indicator of technology being imported and not available within the country. Similarly, in a vast and heterogeneous country like India, there are huge technology disparities within. These gaps need to be filled for the sake of inclusive and optimised growth. These technologies are mostly non-proprietary, unlike those in the case of imports. NSTIS needs to evolve mechanisms for monitoring and dissemination of such technologies. As an example, Indian MSMEs which employ a large fraction of workforce and are big contributors to GDP are very low in productivity by international standards. They do not have access to better technologies and government programmes to support their technology upgradation usually ends in importing machinery, while many of Indian national labs are privy to such advanced technologies, techniques and processes. There could be a matching pair of donors and receptors of advanced techniques and technologies. NSTIS has to organise sector wise scout groups that can identify match making possibilities between these pairs. Lastly, major Indian national labs are repositories of several advanced technologies which do not see light of the market for a variety of reasons. They do have certain mechanisms for transfer of spinoff technologies to private sector. This is an inefficient way of transferring accidental or residual set of technologies to sometimes unworthy hands. NSTIS needs to set up a central technology transfer office, where all such data is collated and wherever possible it may encourage spinning off startups instead of going through rig morale of technology transfer. Scientists, technologists and innovators involved in the development process are the best to take that technology to climax and they should be provided due support by their parent organisation. Thus, unobstructed flow of developed technology may maximise benefits.

Repository of STEM Human Resources and Managing Talent

India is a country which loses hundreds of its young bright students to international ecosystem every year. This is a national loss and India does not have an easy solution to this. Talent is lost because India cannot provide the right kind of jobs, salaries, incentives and ecosystem as compared to developed countries. There are somethings which cannot be controlled, but there are other steps which can salvage a good fraction by right management. In many trades, there is excess of STEM workforce and deficit in another category. NSTIS needs to generate a category wise repository of human resources of STEM workforce in India covering a variety of knowledge base skills (STEM HE, R&D and manufacturing related). With the help of industry, chambers of commerce, national labs and academic institutions, it may have to prepare a comprehensive and dynamic demand scenario for different categories of technicians, maintenance engineers, science teachers, researchers and technologists. It has to prepare rigorous procedures and methodologies to channelise and filter students into various streams according to the demand and evolve strategies to cope with excesses and deficits with help of such repositories. This may require several associated regulatory measures. It may have to emulate the example of OECD group of countries in this context as they guide their member countries based on their robust database. NSTIS may also have to identify a reserve group of professionals and groups with less volatility/mobility like superannuated, women and deprived groups to compensate deficit. It may also have to address issues of excess talent in a sector either by pre-emptive action on the number of seats or taking up new relevant challenges in that sector and involving academic institutions or/and industries. Interventions with the right kind of salaries and ecosystem may also be helpful in attaining such goals. Roping in private greenfield initiatives in pertinent areas may open new possibilities for STEM talent.

Appellate Authority and Ethics Monitor

There are a large number of potential scientific or technology innovations with probable linkage to manufacturing which never see light of the day. There are multitudes of reasons related to bureaucracy, hierarchical restrictions and limitations of a regulatory regime or plain finance. NSTIS needs to set up an appellate authority which can be accessed by any scientific worker, technologist or innovator. Their grievance should be examined on lines similar to the way administrative grievances are handled by the Central Administrative Tribunal. These could come from grassroot workers both in public and private sectors. There are many other malpractices like plagiarism, use of fake journals, credit hogging etc. which also should be addressed by this authority, which can be overseen by NSTIS.

Harnessing Diaspora

There is a huge STEM talent stock of Indian origin working in many foreign countries in high technology areas. With rise of protectionist tendencies in many countries, it is likely that quite a number of them may like to come back to India. This is possible only if the right conditions exist. Some of Indian institutions have been, at their own level, attempting to recruit faculty or researchers in general, but without much success. If this exercise is carried out in a much more systematic way with background preparations, then it may be more fruitful. NSTIS has to organise high powered sector-based teams for recruitment of such talent at different entry levels/positions. The Government of India has recently allowed lateral entry for joint secretary level administrative posts. It should open possibilities for STEM workforce also to be recruited in government labs. Quite often, the candidates need suitable assurances which only a fully authorised recruiter can offer. This may also necessitate handling suitably the potential ruffled feathers in domestic institutions. Such a recruitment mission should aim to fill gaps in specialists of contemporary subjects. This could aim for quality faculty, researchers or even enterprise.

STEM HE, scientific research, technology development and manufacturing sectors seem to be progressing in India as if oblivious of each other. STEM R&D is one of the major pillars that can directly make an impact on high-tech manufacturing. Prof. CNR Rao, former Principal Scientific Adviser to the Prime Minister of India, commented some time back that there is a total disconnect between Indian scientific research and manufacturing. The famed 'Mangalyaan' mission to Mars by ISRO did not carry any payloads from any Indian Institutes of Higher Education. Departments of space and defence find it difficult to scale up their operations despite useful technological developments by these organisations. There could be many more such examples, as mentioned in previous chapters, to prove poor linkages between scientific research in universities with high technology and manufacturing. These are pointers to Indian STEM researchers that they must reinvent themselves for the sake of linking themselves to high technologies and manufacturing industries.

The processes of reformation and reinventing have to go hand in hand. NSTIS has to play a catalytic role in achieving several such goals.

Reforming STEM R&D Ecosystem

There are plenty of ills which affect the R&D ecosystem and limit its potential. Excessive bureaucracy and hierarchy in most STEM R&D institutions are restrictive to both innovation and promoting collaborations with other institutions and industries. Similarly, performance appraisals of scientists, technologists and projects fall short in ensuring objectivity, fairness and transparency. Even peer review mechanisms of individual scientists, technologists or projects are quite distorted and often stage managed. While NSTIS has to frame guidelines for fairness and transparency of the process, the execution mechanisms have to be implemented by institutions themselves. Some of these variables can be suitably incorporated as one of the metrics for ranking STEM R&D institutions. There are many such problem areas described in earlier chapters in greater detail.

New Methods for Dissemination of Knowledge Base

In India as well as many developing countries, knowledge base dissemination between institutions and from institutions to industry or marketplace is quite poor. There is wide range of variation in status of knowledge base and skills (STEM linked) required for a specific task between institutions in the country. Indian STEM R&D ecosystem may have to develop a novel means to fill this gap. Most institutions follow a system of sabbaticals or deputation for their post-doctoral fellows. In the process, scientific workers visit international labs for better exposure and interactions. What is being suggested here is to encourage young scientific workers to go for a system of alternate sabbaticals, where they are encouraged to visit a mass institution of much lower national ranking within the country for the sake of transfer of knowledge base. This could be in the form of setting up a new lab with funding provided for the same by the government, training manpower or a series of lectures, or both. This should be in the nature of handholding between two institutions. The Government of India has recently instituted special fellowships of large amounts for few select institutions. This is aimed to retain talent within the country. Unfortunately, this may not necessarily work as it is a half-hearted measure. Instead, it is required that a seamless connect is established between researchers, institutions and users of the research. Results of such research should further be judged by the user. Such research establishes linkages between researchers and industries and also promotes retention of talent within the country. Similarly, for interaction with small industries, another special mechanism needs to be evolved. Most government R&D institutions with annual funding greater than certain amount need to be incentivised to transfer results of their research to low productivity entities within the country. Indian MSMEs and agriculture are very low in productivity by international standards. R&D institutions could be given excess funding (about 5% of their annual intake) for the sake of such work. They could autonomously decide what kind of support they can provide, but they could be assessed for this work by NSTIS. This funding could be altered on annual basis, depending upon their performance. This may again be a handholding type of outreach. Organisations participating in this exercise may be allowed to recruit some additional manpower on temporary basis in the beginning. Details of modus operandi may be decided by NSTIS in consultation with government and appointed scout mechanisms.

The Government of India has recently decided to give extra push to few selected areas of industrial production in what is called mission 'Make in India.' The choice of these areas was based on market research related to their demand and the potential to incentivise employment. These sectors have been chosen, irrespective of source of technology involved, and its nature. The metric of progress is related to FDI. No attention has been given to encouraging scientific research and technology development related to these areas. It is evident that India faces a major challenge in influencing attitudes and mindsets of policymakers in the context of role of R&D vis-a-vis FDI or offsets in promoting manufacturing. All these sectors and even several others are deserving cases for incentivising scientific research and technology development in relation to industrial manufacturing. But more important is to list down R&D themes which have significant potential bearing to promote hi-tech manufacturing across various sectors. Some of these themes listed below are selected based on their present level of neglect, but do have a significant potential impact. Some are also related to futuristic/disruptive technologies. These themes are expected to build a robust industrial base, remove hurdles to high technology-based manufacturing, carve out niche for Indian technology products and establish intimate links between R&D and manufacturing. These should also encourage inclusive, sustainable and distinctive trajectory of growth of high technology industries promoting quality employment. These themes may help meet the challenges of managing disruptive technologies.

Environment Monitoring and Control

Excessive environmental restrictions have often come in conflict with the goal of industrial development in India. The question remains, what is 'excessive' and who decides it. Environment monitoring is equivalent to 'prevention is better than cure' on a larger scale. Environment assessment is sought both prior to an industrial project as well as after the start of its operation. In India, as industrial development picks up, these tasks will grow in volume, requiring a large amount of air and marine pollution to be monitored. The probability of environmental disasters or accidents may grow. India may need an agile and responsive environment monitoring mechanisms in place. In the absence of these mechanisms, the usual knee jerk response is to ban products or processes or shut down plants, which impedes industrial growth. Such decisions are likely in the absence of a robust data on environment and poor understanding of the science behind the process or inability to provide alternatives. It is a challenge for Indian researchers to develop robust, efficient and economic techniques of environment monitoring and control, replacing existing mechanisms which do have limitations. This should help in pinpointing sharply the sources of environmental troubles. They have to evolve a methodology based on these techniques as well as scientific analysis of data generated that only polluter pays and not the whole society or its industrial development. There are several other aspects of environment monitoring and control which call for innovative solutions and technologies. The theme demands scientific research, technology development at levels of pollution source, monitoring and control.

Quality Monitoring, Control, Safety and Reliability

All these issues related to quality monitoring (of all manufactured products), safety and reliability have got poor attention both from consumers as well as manufacturers. This is intimately related to economics of the processes as well as the inability to come out of tradition. This mind-set percolates even to the exporters that are thus not able to compete internationally. In principle, such abilities to produce quality goods does exist in India as is demonstrated by its strategic sectors like atomic energy and space for whom these are all imperatives. Thus, the problem of a large range of quality products is related to market economics. The challenge for researchers and innovators is to develop techniques and produce quality monitoring instruments and products which are economically acceptable. While China, in general, produces cheap but low quality goods, manufacturers in developed countries are known to produce quality goods but at a higher cost. Indian researchers can in principle aim at combining quality with low cost — first to serve Indian consumers and then to influence exports. What is true for quality is equally true for safety and reliability. These are background themes for most sectors, but have the potential to impact the ecosystem of manufacturing significantly.

Components and Products

A serious lacuna in Indian industrial development in general is related to what is called industrial base. Most of scientific research, technology development as well as manufacturing in India (both public and private sector) give thrust to building products and big systems, but neglects small components, ingredients, devices and materials. India's strategic sectors are no exception. Indian public sector manufacturers, including those in defence, first build systems and then spend decades improving upon the percentage of indigenisation. Most of such indigenisation is not of critical components. This kind of neglect is true in the case of fighter aircraft, tanks, pharma products, solar power systems, foundries for chips or precision machine tools etc. There can be several more examples. These could prove costly in terms of lifecycle costs or even critical dependence on foreign sources. These efforts are neglected as they may be more challenging, require huge capital or offer less demand/returns. A greater emphasis on scientific research related to developing components, devices and materials can be a boon for the nation in several ways. More significantly, it could lead to unleashing the creativity of its innovators. It could also persuade more hi-tech manufacturers to invest in the country as hi-tech components can be sourced locally.

Advanced Manufacturing

Developing countries like India are likely to be confronted with challenge of several disruptive technologies. These could be like robotics, auto-vehicles, IOT, additive manufacturing etc. Most of them have the potential to disrupt the employment scenario by altering both internal as well as external demand. Most of the time, a new technology emerges in the developed world and developing countries follow it with some time lag. In case of these disruptive technologies, it cannot be the same pattern as the responses of developed and developing countries may diverge because of different nature of impact. This poses a challenge to researchers and technology developers in developing countries. They may have to find innovative applications or mode of deployment of these technologies (some of these have been discussed in earlier chapters). This may require imagination and nurturing the right kind of human resources on a large scale. India has launched a special programme in its schools by the name 'Atal Innovation Mission' on some of these specialised technologies. Presently, it is limited to a very small number of Indian students and it needs to be spread to remote areas. Its widespread dispersal may require broadband communication network. India does have such a programme called BHARATNET and is rolling out 5G networks. This needs to be expedited as speedy adaptation to these technologies is the key to facing some of these challenges.

Efficient Utilisation of Resources

Indian society in general has traditions and ethos of efficient utilisation of resources, but with onward march of influences from the western culture based on 'one time use' technologies as well as increasing volumes of all types of waste, the problems related to waste management have become more challenging. This has affected environment and has decreased the efficiency of utilisation of resources. Efficiency in utilisation of all types of resources is imperative as development may cause scarcity of most natural resources ranging from water to rare earths or atomic minerals. These problems are global in nature and poor countries like India can ill afford such practices. It is for innovators, scientists and technologists to find alternate means to ensure better waste (all kinds) management and efficient utilisation of resources. This also implies development of technologies for best value addition to all raw resources. It requires a holistic thought on part of designers and innovators and may require efficient design of products and processes to facilitate retrieval of precious parts and components from used products. The theme of efficient utilisation ranges from sectors like mining and minerals, agricultural products and waste, electronic waste or even water conservation.

Small is Beautiful

This is originally a Gandhian thought which was also endorsed by some economists in the context of developing rural based economies. The concept may not be universally applicable, but it does have relevance today in many situations of technology deployment. Scientists and technologists in developing countries have to innovate to find distinctive and unique modes of technology deployment. This may be particularly necessary in the case of new and disruptive technologies. These could be, for example, innovating and upgrading rooftop solar, mini-hydro power, micro-irrigation, water management, compact and portable quality monitoring instruments (for various products) and mobile medical diagnostic kits etc. These are easier to develop, deploy and manage in case of developing countries. A particular example for the case of potentially disruptive technology is that of additive manufacturing. While NASA imagines it to be used for futuristic space stations, Indian innovators could think of its applications for civilian structure in remote or not so easily accessible areas. Artificial intelligence, which is proposed for economic and labour saving means of production in the developed world, could be used to help make decisions to end human discretion-based corruption in developing societies and other such applications. Finding such unique or distinctive deployment of technology may enrich even global innovation ecosystem.

A large gamut of themes and issues of significance to hi-tech manufacturing have been touched upon in brief. These cannot be isolated from techno-socio-economic realities of the country. India is predominantly an agricultural country and is making a gradual transition towards manufacturing and services. India's service sector is growing much more rapidly as compared to manufacturing and the pattern is distinctly different from that of many of the developed countries. Growth of hi-tech manufacturing is even slower. Employment of its burgeoning young population has to be central to any of its growth strategy. Presently, the Indian establishment is focussed on economic and regulatory reforms to facilitate its traditional manufacturing sector. There are limitations to this approach in the form of socio-political hurdles. The situation is likely to be aggravated by global developments related to several disruptive technologies. Low skill workforce switching from agriculture may not be able to cope with the situation. Prof. Mohammad Yunus (Nobel Prize, Economics) rightly calls technologies like AI and robotics dangerous. There is an associated hazard that most developing countries tend to ape developed countries in technologies they develop or deploy. Onward march of technologies cannot be arrested, but can be countered by being innovative in deployment of these technologies and training its large workforce suitably.

How well prepared is the Indian ecosystem to face these challenges? The principle players that matter are STEM HE institutions, government national labs, industrial manufacturers (both public and private) and private industrial research labs and many of the young startups. Basic infrastructure is vast but limiting factors are their quality, orientation and lack of coordinated response. The network of its government labs spends one of the largest fractional expenditure of total national R&D expenditure as compared to that of most big countries. Many of these labs like departments of space, agriculture, defence, atomic energy and CSIR have several worthwhile achievements to their credit, but have only nominal direct contributions related to manufacturing. India does have few islands of STEM HE and R&D, but its mass institutions are highly impoverished in terms of support (fraction of total national expenditure on R&D) and consequently R&D output. Except for some of the elite institutions, other university departments neither take cognisance nor are competent for industrial R&D. Most of its private industries are indifferent to R&D. Few of the sectors where Indian industries have invested in R&D like pharmaceuticals, automobiles and IT, have been able to make a global impact. In latest Forbes list of top 2500 R&D spenders, there are only 26 Indian industries and all belong to these sectors. Manufacturers in the government sector have mostly relied on technology based on international licenses. India has a large number of unregistered MSMEs with low productivity and poor exposure to R&D.

As it seems, Indian society as well as policymakers have relied on scientific research and technology development more for strategic goals and less as an instrument of economic development. Indian talent and basic infrastructure for R&D is harnessed sub optimally. The proposed NSTIS has to bring about a coordinated response from all relevant players to the challenge of bringing about technology-led growth. The concerned ecosystem operates at three levels — grassroots level researchers and technology developers, organisational leadership and policymakers. Indian researchers have performed very well at individual levels in a variety of situations. As Ross Basset, the author of *Technological Indian,* mentions in his book, a large number of Indian technologists have contributed immensely to the global technology ecosystem in recent times, but they have not benefitted India's growth. NSTIS has to stimulate formulation of relevant policies and ensure consequent actions related to second and third performance levels. These should be aimed at building up symbiosis that impacts hi-tech manufacturing based on indigenous R&D. Some of the foundational steps for this purpose are treating STEM HE on par with other national strategic objectives, developing potential to attract and retain talent in face of competition from developed countries, unhindered movement of talent across institutions and industries when and where required. Undiluted competition at all levels in STEM HE and reconciling to affirmative action by encouraging intra-group competition. Affirmative action should be used as a tool to widen search for talent. NSTIS need to ensure that universities synchronise their R&D directions with contemporary technology needs, while a fair balance is attained in support to elite and mass institutions. The swathe of policies and actions required is very wide and these points form only the basis. More significantly, NSTIS should spearhead these reforms rather than being a toothless body. There may be necessity of several diverse entities to join hands for accomplishing these goals. NSTIS could go one step further to attract talent from other developing countries with commonality of perspectives to work out distinctive solutions to common problems.

There is enough talent within India which is presently harnessed sub optimally. To retain it within, there needs to be a well-planned strategy to match supply and demand as well as assurances on fair and genuine respect for talent and innovation. Finally, India has to nurture and value its human resources to meet the challenges of disruptive technologies. One may recall the prophetic words of John F Kennedy, who said, “We believe that if men have the talent to invent machines that put men out of work, they have the talent to put those men back to work.”

REFERENCES

[1] OECD, *"OECD Science, Technology and Industry Scoreboard 2015: Innovation for growth and society"*, OECD Publishing, Paris. ISBN 978-92-64-23978-4 (PDF) (2015) http://dx.doi.org/10.1787/sti_scoreboard-2015-en.

[2] OECD, *"New Sources of Growth: Knowledge-Based Capital Key Analyses and Policy Conclusions – Synthesis Report"* (2013) https://www.oecd.org/sti/inno/knowledge-based-capital-synthesis.pdf.

[3] Redden Elizabeth, *"An End to Years of Growth for New International Enrollments"* (2017) https://www.insidehighered.com/admissions/article/2017/11/13/after-years-growth-us-colleges-see-decline-number-new-international.

[4] Krugman Paul, "A Model of Innovation, Technology Transfer, and the World Distribution of Income." *The Journal of Political Economy* 87: 253-266 (1979) https://www.jstor.org/stable/1832086.

[5] *Research & Development statistics at a glance,* 2017-18, Department of Science & Technology, Ministry of Science and Technology, Government of India, New Delhi. http://www.nstmis-dst.org/Statistics-Glance-2017-18.pdf.

[6] FICCI, Higher Education Summit, "*Higher Education in India: Moving Towards Global Relevance & Competitiveness."* (2014) http://ficci.in/spdocument/20513/FICCI-EYReport2014.pdf.

[7] Planning Commission, Govt. of India "*The Manufacturing Plan, Strategies for accelerating Growth of Manufacturing in India in the 12th Five Year Plan and Beyond*" http://www.planningcommission.nic.in/aboutus/committee/strgrp12/str_manu2703.pdf.

[8] Dutz Mark A. "*Unleashing India's innovation : toward sustainable and inclusive growth*" World Bank, Washington D.C.20433, ISBN 978-0-8213-7197, ISBN 978-0-8213-7198-5 (electronic) (2007) http://siteresources.worldbank.org/SOUTHASIAEXT/Resources/223546-1181699473021/ 3876782-1191373775504/ indiainnovationfull.pdf.

[9] KPMG, "*The New Wave Indian MSME: An Action Agenda for Growth,*" (2015) https://assets.kpmg.com/content/dam/kpmg/pdf/2016/03/The-new-wave-Indian-MSME.pdf.

[10] *India Brand Equity Foundation*, Department of Commerce, Ministry of Commerce and Industry, Government of India. https://www.ibef.org/industry.aspx.

[11] Gupta Vaibhav, Ghosh Arunabha. "*Be resource efficient to make in India*" www.business-standard.com/article/opinion/vaibhav-gupta-arunabha-ghosh-be-resource-efficient-to-make-in-india-115091401272_1.html Last Updated on September 14, 2015.

[12] United Nations, Department of Economic and Social Affairs, Population Division "*World Urbanization Prospects: The 2018 Revision*" (2018) https://population.un.org/wup/Country-Profiles/.

[13] IBEF report, "*Indian pharma sector surges in global market*" (2015) http://www.pharmabiz.com/NewsDetails.aspx?aid=90553&sid=2.

[14] Tellis Ashley J., Raja Mohan, "*The Strategic Rationale for Deeper U.S.-Indian Economic Ties*" (2015) http://carnegieendowment.org/2015/ 08/ 07/ strategic-rationale-for-deeper-u.s.-indian-economic-ties-pub-60922.

[15] IHS Market, "*Jane's Defense Industry*" (2017) https://www.janes.com/images/assets/755/69755/Briefing_South_Korea_prepares_offset_reforms.pdf.

[16] World Health Organization, "*Seven million Premature Deaths Annually*" (2014) http://www.who.int/phe/eNews_63.pdf.

[17] Behera Laxman Kumar. IDSA Monograph Series, No. 21 "Indian defence industry, *Issues of Self Reliance"*, ISBN: 978-93-82169-21-5 (2013) https://idsa.in/system/files/monograph21.pdf.

[18] Lal H., FICCI, *Global R&D Summit, Destination India*, July 25-26, 2013 The Ashok, New Delhi, http://ficci.in/spdocument/20284/FICCI-Battelle-Knowledge-Paper-Global-R&D-Summit-2013[1].pdf

[19] Greenemeier Larry, Scientific American "*NASA Plans for 3-D Printing Rocket Engine Parts Could Boost Larger Manufacturing Trend"* (2012) http://www.scientificamerican.com/article.cfm?id=nasa-3-d-printing-sls-rocket-engine.

[20] Andersson Eric, SITC Bulletin Analysis May *2013 "Additive Manufacturing in China: Aviation and Aerospace Application (Part 2)"* (2013) https://escholarship.org/uc/item/7h12120m.

[21] Ishengoma Fredric R, Mtaho Adam B. "3D Printing: Developing Countries Perspectives" *International Journal of Computer Applications* 104 No 11: 0975 – 8887 (2014) https://pdfs.semanticscholar.org/c251/6eba4c2c1d3254ac33c413d52f7ea79f8be4.pdf.

[22] Bhabha, H.J., *Last Lecture delivered by Invitation of International Council of Scientific Union on 7 January, 1966* (private information) http://www.tifr.res.in/~hbbccc/download/dlal.pdf.

[23] Balance, R., Pogany, J. and Forstner H. *"The World's Pharmaceutical Industries: An International Comparative Study"* Aldershot: Edward Elgar (1992).

[24] *Rising Sun Disruption on the Horizon*, November 2015 https://home.kpmg.com/in/en/home/insights/2015/11/therisingsun-disruptiononthehorizon.html.

[25] FICCI-Battelle, *"India's emerging competitiveness as Global R&D Destination India, Knowledge paper"* July 25-26, 2013 The Ashok, New Delhi (2013) http://ficci.in/spdocument/20284/FICCI-Battelle-Knowledge-Paper-Global-R&D-Summit-2013[1].pdf.

[26] Testimony before the US-China Economic and Security Review Commission Hearing on March 30, 2011 One Hundred Twelfth Congress "*Chinese State-Owned Enterprises and US-China Bilateral*

Investment" https://www.uscc.gov/sites/default/files/transcripts/3.30.11HearingTranscript.pdf.

[27] Evenson R. E., Pray Carl E., Rosegrant Mark W. *"Agricultural Research and Productivity in India"* Research Report 109, International Food Policy Research Institute, Washington DC (1999) http://www.indiaenvironmentportal.org.in/files/rr109.pdf.

[28] *India Brand equity Foundation IBEF*, Science and Technology, October 2017 https://www.ibef.org/download/Science-and-Technology-October-20171.pdf.

[29] Keenan Michael *"Moving to the Innovation Frontier: Lessons from the OECD"* Chapter 2 of Korean Science and Technology in an international perspective (2012) https://www.springer.com/us/book/9783790827521.

[30] Lee Kong-Rae, University Industry Collaboration in Korea's National Innovation System, *Science, Technology & Society* 19:1–25 (2014) https://doi.org/10.1177/0971721813514262.

[31] NASSCOM Study *"Indian Startup Eco System -Traversing the Maturity Cycle"* (2017) https://www.nasscom.in/knowledge-center/publications/indian-start-ecosystem-%E2%80%93-traversing-maturity-cycle-edition-2017.

[32] Report of Joint Committee of Industry & Government (JCIG) *"White Paper on Stimulation of Investment of Private Sector into Research & Development in India"* Department of Science & Technology, Ministry of Science & Technology, May 2013. http://citeseerx.ist.psu.edu/viewdoc/download?doi=10.1.1.694.4211&rep=rep1&type=pdf.

[33] Jung Jiwon and Mah Jai S. "R&D Policies of Korea and Their Implications for Developing Countries" *Science, Technology & Society* 18:165–188 (2013) http://journals.sagepub.com/doi/pdf/10.1177/0971721813489435.

[34] Jung Heyjie and Mah Jai S. "The Role of the Government in Science and Technology Education of Korea", *Science, Technology & Society* 19:199–227 (2014) https://doi.org/10.1177/0971721814529877.

[35] Kaul Sanat Indian *"Higher Education in India: Seizing the Opportunity"* Council for research on International Economic Relations Working Paper No 179 (2006) http://icrier.org/pdf/WP_179.pdf.

[36] Carnoy Martin et al. *"University Expansion in changing Global Economy: Triumph of BRICS"* Stanford University Press ISBN: 9780804786010, Digital ISBN: 9780804786416 (2013) https://www.sup.org/books/title/?id=22662.

[37] OECD Report *"Knowledge Based Economy"* (1996) https://www.oecd.org/sti/sci-tech/1913021.pdf.

[38] FICCI Higher Education Summit "*Private sector participation in Indian Higher Education*" (2011) https://www.ey.com/Publication/vwLUAssets/Private_sector_participation_in_Indian_higher_education/$FILE/Private_sector_participation_in_Indian_higher_education.pdf.

[39] FICCI Higher Education Summit *"Higher Education in India: Vision 2030"* (2013) https://www.ey.com/Publication/vwLUAssets/Higher-education-in-India-Vision-2030/ $FILE/ EY-Higher-education-in-India-Vision-2030.pdf.

[40] Xing Yuqing, Dumon Jean-Christophe, Baruah Nilim, *"Labor Migration, Skills & Student Mobility in Asia"* (2014) https://www.oecd.org/ migration/ Labour-migration-skills-student-mobility-in-Asia.pdf.

[41] U.S.-China Economic and Security Review Commission "*Report to congress of the U.S.-China Economic and Security Review Commission"* US Government Printing office Washington, (2011) https://www.uscc.gov/sites/default/files/annual_reports/annual_report_full_11.pdf.

[42] Barret Edward Clark, *China Brief* Volume: 14 Issue: 24, Chinese servers China's "Server Sinification" Campaign for Import Substitution: Strategy and Snowden (Part 1) (2014).

[43] https://jamestown.org/program/chinas-server-sinification-campaign-for-import-substitution-strategy-and-snowden-part-1/.

[44] Feynman. R. "What is science" *Physics Teacher*, Vol.7: 6, 313 (1968) http://profizgl.lu.lv/pluginfile.php/32795/mod_resource/content/0/WHAT_IS_SCIENCE_by_R.Feynman_1966.pdf.

[45] OECD *Science, Technology & Industry Outlook*. OECD Publishing (PDF) Page 312 ISBN 978-92-64-17032-2 (print) ISBN 978-92-64-17039-1 (2012) http://dx.doi.org/10.1787/sti_outlook-2012-en.

[46] National Science Board. Science and Engineering Indicators *"Chapter 6, Industry, Technology & Global Marketplace"*, Alexandria, VA, National Science Foundation (NSB-2018-1), (2018) https://nsf.gov/statistics/2018/nsb20181/assets/nsb20181.pdf.

[47] Basset Ross. Harvard University Press *"Technological Indian"* (2016) https://www.amazon.in/Technological-Indian-Ross-Bassett/dp/0674504712.

About the Author

B. M. Suri, PhD

B. M Suri, PhD is a physicist by profession (post graduation with distinction from University of Delhi, Doctorate Mumbai University). His work spanned specific areas related to laser science and technology. After more than three decades of pioneering contributions to several applications relevant to advanced nuclear technology, he superannuated from Bhabha Atomic Research Centre, Mumbai, India. Worked in international laboratories on varied subjects of both basic and applied interest. For the sake of dissemination of knowledgebase in his working field, organized several national and international conferences as well as schools for young doctoral and postdoctoral scholars. He has mentored and guided several students for their post graduate dissertations and PhD under auspices of Mumbai University and Homi Bhabha National Institute of Education. He had been a Professor of Physical Sciences at Homi Bhabha National Institute for Education, Mumbai and served in advisory capacity in several councils of Department of Atomic Energy, Defence Research Labs and Department of Science & Technology in India.

Presently, post superannuation, Suri's current interests are to apply scientific and technological developments for several applications and continues to serve various organisations in advisory capacity. His main interests being those related to Environment, Renewable Energy and Advanced Manufacturing.

INDEX

A

B

C

D

E

K

L

M

N

O

P

Q

R

S

T

Imperial Maladies: Literatures on Healthcare and Psychoanalysis in India

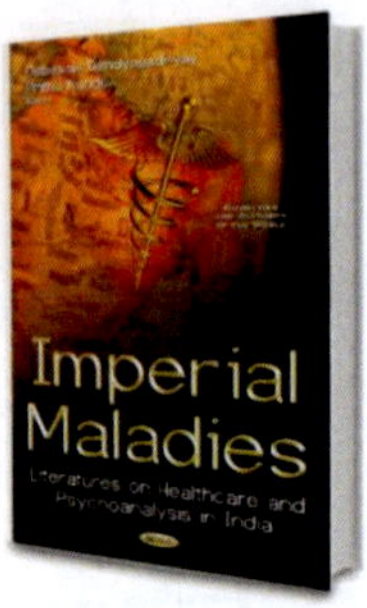

Editors: Pritha Kundu and Debashis Bandyopadhyay

Series: Countries and Cultures of the World

Book Description: The thrust-area of this book is the connection between imperial anxieties and tropical health situations along with intriguing psychological questions involving race, politics, gender, history and colonial modernity.

Hardcover ISBN: 978-1-53611-863-6
Retail Price: $195

Reports on British Prison-Camps in India and Burma

Author: Oscar Moten

Series: Political Science and History

Book Description: This book is a compilation of the reports by the International Red Cross Committee on the British Prison-Camps in India and Burma in February, March and April 1917. The British camps in India and Burma for Turkish prisoners of war and civil residents in the Indian Empire of enemy nationality were visited by three accredited representatives of the Red Cross.

eBook ISBN: 978-1-53613-778-1
Retail Price: $82